A, B, SEA

Also by Jack Lagan: *The Barefoot Navigator: Navigating with the Skills of the Ancients*

Praise for *The Barefoot Navigator*

"*The Barefoot Navigator* is a cleverly written little book that combines the history of the ancient art of navigation with the practical application of those techniques today." —*Ocean Navigator*

"From how to analyze clouds and currents to determine direction to how 21st-century sailors can integrate these techniques with modern equipment, any sailor will find *The Barefoot Navigator* intriguing." —*Midwest Book Review*

"An exciting read that makes a compelling argument: To navigate better and with greater satisfaction, sailors must learn more about the sea around them."—*Cruising World*

"This book investigates . . . navigation capabilities and shows you how practical, technology-free navigation can be used to enhance modern navigation. Interesting stuff!" —*Latitudes & Attitudes*

A, B, SEA

A LOOSE-FOOTED LEXICON

second edition

Jack Lagan

SHERIDAN HOUSE

Published by Sheridan House
4501 Forbes Boulevard, Suite 200, Lanham, Maryland 20706
www.rowman.com

10 Thornbury Road, Plymouth PL6 7PP, United Kingdom

Distributed by NATIONAL BOOK NETWORK

British Library Cataloguing in Publication Information Available

Library of Congress Cataloging-in-Publication Data

Lagan, Jack.
A B Sea : a loose-footed lexicon / Jack Lagan.
p. cm.
Includes bibliographical references.
1. Sailors—Language—Dictionaries. 2. English language—Slang—Dictionaries.
3. Naval art and science—Dictionaries. I. Lagan, Jack. II. Title.
PE3727.S3 L34 2003 2003000286

ISBN 978-1-57409-322-3 (pbk. : alk. paper)
ISBN 978-1-57409-323-0 (electronic)

∞™ The paper used in this publication meets the minimum requirements of American National Standard for Information Sciences—Permanence of Paper for Printed Library Materials, ANSI/NISO Z39.48-1992.

Printed in the United States of America

TEN REMARKABLE FACTS FROM THE LOOSE-FOOTED LEXICON

1. There are so many names for the different parts of a boat that even professional seamen would forget them. When such unfortunate lapses of memory occurred, they would substitute "chicken-fixing," "gill-guy," "wim-wom," or even "gadget." "Jack! Haul in that er . . . timmey-noggy!" See GADGET.
2. The principle of the "diesel" engine was first drawn up in 1890 by an Englishman with the splendid name of Herbert Akroyd. If history had run a different course, we would be calling on the helmsman to "Start the Akroyd!" See DIESEL ENGINES.
3. It wasn't until three years later that a German engineer called Rudolf Diesel patented the diesel engine as we now know it. Diesel died in mysterious circumstances in 1913 when he disappeared over the side of a ferry; Akroyd was not a suspect. See DIESEL ENGINES.
4. A GPS receiver once appeared as a key witness in a murder trial. See MURDER ON THE HIGH SEAS.
5. Satellite navigation systems such as GPS are the only known practical application of Einstein's Theory of Relativity. See GPS, GALILEO.
6. Seamen in the days of sail were allowed to wear a ring through one ear to show they had passed the Cape of Good Hope and in the other after passing Cape Horn. See EARRING. Similarly, officers were allowed to put a foot on the Wardroom table for each Cape passed.
7. To seduce the maggots and weevils out of their hard tack biscuits, sailors would cover a plate of them with fresh fish entrails. See HARD TACK.

8. The 100 percent guaranteed instant cure for seasickness is to stand under a tree. See **SEASICKNESS**.
9. In the days long before Emergency Position Indicating Radio Beacons (EPIRBs) there was an understanding among seamen that, should they run out of food while adrift at sea, they would start to eat each other. See **CANNIBALISM**.
10. The biggest cocktail *ever* was made on October 25, 1599, by Sir Edward Kennel, Commander in Chief of the English navy. The monster punch—enough for 6,000 party guests—included eighty barrels of brandy, nine of water, 25,000 large limes, eighty pints of lemon juice, 1,300lb of Lisbon sugar, 5lb of nutmeg, and 300 biscuits, plus a giant cask of Malaga wine. Could Sir Edward have overdone it with the water? See **PUNCH**.

INTRODUCTION

Throughout the course of the 20th century, and especially after World War II, seafaring became more of a personal pursuit and less of an industry; more civilian and less military; and at the same time more high-tech and less rudimentary. But the underlying elements have remained the same: boats, sea, wind, and seafarers in search of excitement, adventure, and escape. For much of the latter half of that century I have raced dinghies, fished offshore, delivered charter-yachts, fixed weather-fax systems, gunk-holed in skiffs, and lived aboard a junk-rigged schooner—anything to avoid having to turn up at an office before my eyes could make out shapes.

Words can develop special meanings for each individual; they can trigger off memories, emotions, ideas, and even opinions. The selection of terms and expressions for inclusion in this lexicon was based purely on personal preference and influenced by my fascination, as a writer and navigator, with the origin of words in everyday use in today's seafaring community. English is a great borrower, and side by side with words that stem from Old English and Anglo-Saxon are just as many that come from Dutch, French, Old German, Norse, Spanish, Tamil, Malay, and even Chinese. The genealogy of jargon is global.

It is fascinating that words written in the annals of 14th-, 15th-, and 16th-century voyaging remain in the vernacular of 21st-century seafarers—but it is far from thrilling if you do not know which ones they are. For that reason I have included illustrative passages from classic books of the sea. These sources include some novels and short stories; when Jack London describes a boat, you can be confident that the words he uses, even in fiction, are technically the correct ones. In other cases the entries here are an unabashed vehicle for a personal anecdote or for a little tub-thumping.

Inevitably for a book with a jib cut like this one's, I was drawn into an exploration of those nautical expressions that richly texture everyday English. We are predisposed, wrongly in my view, to condemn out of hand the use of jargon. The improvised invention of specialist terminology by groups with common interests is justified by their need for a shorthand form of communication. This is happening right now in the burgeoning world of computers and the Internet. As personal computers and the World Wide Web become increasingly influential in economic and social spheres, the language spills over into everyday parlance; people "download" ringtones for their "user-friendly" cell phones, which they use to "network" socially.

The influence of the jargon of the sea on the world's most important and adaptable language—*pardonnez-moi, mes amis français*—is hitherto unparalleled. Life in seagoing nations has over centuries led to the absorption of hundreds of borrowed expressions whose users are mainly clueless as to their origin. America, Canada, Australia, New Zealand, and the English-speakers of South Africa have inherited and enriched the legacy, as have those nations of the developing world on whom English was imposed as a consequence of Britain's 18th- and 19th-century sea-borne imperial strategy. By and large, therefore, it has been unnecessary to plumb the depths in order to trawl up some lively specimens.

It is night on the open sea and you are on watch at the helm. Every few minutes or so you dutifully perform a disciplined 360-degree scan of the featureless ocean around you. The rest of the time, apart from an occasional glance at the compass, you are looking up at the breathtaking sky. In tropical climes, away from the polluted atmosphere of so much of North America and Europe, the night sky looks three-dimensional, its distances measured in eternities and putting the lie to the navigator's arbitrary convention of the celestial sphere. You glance at Orion, the most magnificent of all constellations, now just above the crosstrees on the starboard side, and you know that it is roughly 23 hours local time.

It was not by fate that the seafaring community was the first to embrace the Global Positioning System. It has always been so with technology: the magnetic compass, the astrolabe, the quadrant, then the sextant; fine precision instruments that no serious navigator rejected because they were new-fangled technology. Harrison's dogged pursuit of the accurate ship's chronometer solved the problem of longitude. And the very first computer, Babbage's mechanical Analytical Engine was conceived in an era when good navigation tables were the foundations on which empires were built. Fast-forward a hundred years to the early 1960s and I am standing in the wheelhouse of a Cornish long-liner staring at the digital readout on a Decca Navigator. It is so unsophisticated that it is incapable of converting the position to latitude and longitude; the arbitrary numbers have to be plotted on special charts, but it is good enough to put us right over a wreck stuffed with *conger oceanicus*—monster eels that demand a monster rod and reel.

Fast-forward another thirty-five years and I am sitting at the navigation table of a yacht in the Islands staring at a digital chart on the color screen of a powerful laptop computer. I click on "zoom" and I can see a little red boat marking our position on the top of a condominium overlooking the marina. Well, I think, thirty meters off is something I can live with. Having said that, should I be more impressed by being able to look up at Orion off the starboard crosstrees and know that next-on-watch should be brewing my hot chocolate? It would be a tragedy if no one could "read" the heavens anymore and I hope that *A, B, Sea* encourages someone, somewhere to play a part in keeping the heritage alive.

When you step on a boat and nose out through the cut, steering a course for somewhere you have not ventured before, or maybe just heading around the point, you are participating in a drama that started centuries ago and is not yet into the third act. The sea remains the Great Challenge.

May there always be a hand of water under your keel.

Jack Lagan
October 2002

PREFACE TO THE SECOND EDITION

In the ten years since I wrote the first edition of *A, B, Sea* there has been a lot of traveling but nothing like enough of it on the water. Time spent looking at the sea and writing about seafaring is little compensation. But the opportunity to prepare a new edition of this loose-footed lexicon was most welcome.

When I arrived in North Florida in the 1990s to start work on my 55ft junk-rigged Colvin Gazelle, I had thirty-four books with me. They included books about sailing, boat-building, and navigation. After retrieving a few technical tomes, the rest were stowed on a dockside pallet under a tarp: a problem to be tackled later.

The steel hull had been partially fitted out by the owner, a most accomplished ship's carpenter. Sadly, he was a less-than-accomplished electrician and systems engineer. When I switched on the reading light over my berth in the fore-cabin, the STEAMING LIGHT came on. (See also INVERTER.)

Four months later, the electrics had been rewired, furniture had been added to the main cabin, a flash instrument panel to the cockpit and the installation of a hydraulic steering system, which relegated the mechanical one we already had to the role of backup.

Then came the problem of the books. We already had a shelf that ran along the back of the sofa. It had a FIDDLE that ran along the edge but I had a sick feeling in the pit of my stomach: it wasn't going to do the job. Something more substantial was needed but we had to get out of that damn boatyard and start traveling. I improvised with bungee cord. That didn't work either; the sight of all those books heaped on the cabin sole after ten minutes in a swell is too awful to relate here.

So, an important innovation since the first edition is the advent of the e-book. All those books would have been much safer in an e-book reader or a tablet (for better legibility of schematics). Maybe you are reading this on a Kindle? Or on your smartphone?

So why did the previous owner of the Gazelle decide to sell her and give up on all his hard work and fine craftsmanship? After he and his wife took the boat out for the first short sea-trial, his wife decided she was going to hate sailing. They now live in a cabin in Montana.

Jack Lagan
February 2014

This book is not intended to be read from cover to cover, it is meant to be *explored*.

Throughout the text words defined elsewhere are in SMALL CAPITALS.

This book is for my great friend Nick, who died shortly after the publication of the first edition. He probably remains convinced that Gillingham is more than ninety miles from the sea. Nick, nowhere in the British Isles is more than ninety miles from the sea.

aback

1. when the wind strikes the low pressure side (or back) of a sail it is said to be "aback"; technique used when HEAVING-TO, also the sad consequence of failing to free the windward SHEET of a sail when GOING ABOUT.
2. you are "taken aback" when a sudden SQUALL or wind-shift causes the sails to be backed (in other words the crew is claiming it is not their fault); now used in everyday English to mean "taken by surprise."

abaft

toward the STERN or rear of a vessel.

abandon ship

the last thing you want to hear—or think about—when in trouble; probably the most difficult decision any seafarer will have to make.

Your beloved boat can no longer provide security, but leaving too soon could significantly reduce your chances of being rescued. The best advice seems to be: assume the worst. Send a MAYDAY. Launch the LIFERAFT and load it with your emergency GRAB-BAG, WATER, and the rest of the crew. Stay on the radio and continually assess the situation. What is the ETA of the local COAST GUARD or LIFEBOAT search-and-rescue service? Can you get the TENDER into the water too? Is there another boat in the area aware of your situation? If by now you are knee-deep in water, it is probably time to pat her on the COAMING and leave. If the problem is that you are on FIRE, *any* delay could be fatal to you and whoever is waiting in the liferaft. Go!

abeam, on the beam, abreast

an object or mark at right angles to the CENTER-LINE of a vessel is said to be abeam, abreast, or on the beam.

about, going about

the act of going from one tack to the other through the eye of the wind; see TACKING.

abyss

any part of the ocean deeper than 300 fathoms (about 600 meters)

accommodation

the living space on a VESSEL.

If it is better to travel hopefully than to arrive, it is even better to travel comfortably. I would draw the line at chintz curtains, but especially if you are going to LIVE ABOARD, maximizing the comfort of the living space should be a priority. Yacht designers seem to have a preoccupation with maximizing the apparent number of berths instead. This is probably to keep charter companies happy. But if you are sailing solo or short-handed, do you really need all those berths in the saloon? Why not convert a berth into one or two well-upholstered easy chairs? A nice, comfy armchair with a small table to one side for your drink and a navigation repeater to the other so you can keep an eye on the course. So long as you still have a pilot berth for cat-napping in oilies, plus a "proper berth" for when you can get a full night's sleep, then you must have the space.

a-cockbill (pron. "a-co'b'll")

a BOWER ANCHOR hanging free, ready to be let go, is said to be "a-cockbill."

advection fog

a type of fog that occurs when humid (warm, moist) air comes into contact with water cool enough to lower the air temperature below its dew point, thus forcing visible droplets of water to separate from the air and form a fog. The coast of California is very susceptible to advection fog. See CONVECTION FOG.

aft

the rear end of a vessel; the STERN. As a noun: "I am going aft." As an adjective: toward the STERN, "the aft cabin."

aground

not floating; in other words your TIDEFALL calculation was wrong or you were not paying attention to the DEPTH FINDER and, before long, your KEEL is stuck firmly on the BOTTOM and you are brewing a cup of tea.

ahead

forward, in the direction of the BOW, the opposite of ASTERN. On a sailing boat, of course, this is not necessarily the direction in which you are traveling; see HEADING.

a-hull

you are lying a-hull when you have no sails set (see BEAUFORT SCALE); this is usually in a GALE and feeling pretty miserable.

AIDS

no, not navigation aids—the other sort. If you are both footloose and traveling then you are the perfect candidate for victimhood. There are only two remedies for this: abstinence and a good stock of condoms. This advice is particularly directed at women, the only crew members who *really* know whether the condoms are going to be needed.

aid to navigation

any device such as a BUOY or BEACON on land or sea designed to assist in the navigation of a vessel, visually, audibly, or by means of a radio transmission. It is important not to confuse an aid to navigation with a NAVIGATIONAL AID.

Aladdin cleat (Aladdin hook)

a cleat or hook fixed to the BACKSTAY (or main boom) and used to hang a LANTERN over the cockpit; can be very atmospheric. The lamp can also double as an ANCHOR LIGHT, even though it should be in the BOW.

albatross

the large, white, long-range seabird in the family *Diomedeidae*; the name started to appear late 1600s and seems to derive from the Latin *albus*, "white," and the Spanish *Alcatraz*, which in turn is from the Arabic *al-gattas*, "the diver." This seems a little odd because this 10kg (22lb) cross between a duck and a sail-plane is believed to be gatherer of food from the surface of the ocean rather than beneath it. It eats fish, as you would expect, but a good deal of this is ripped from large floating carcasses, suggesting that the albatross is not as nimble as the gannet. In any case, *alcatraz* is the Spanish word for a gannet.

I have never knowingly seen an albatross, but there could be two good reasons for this. Most albatross species live in sub-Antarctic waters, a place that is far too cold for the Lagan species. Also, it is estimated that the world population is less than 10,000. However there are about 1,000 of the Waved Albatross and they live on the Galapagos island of *Española* right on the equator, so there is hope for me yet. Some albatross live as far north as Monterey Bay, California. I have been to Monterey, but did not see any of these magnificent birds from where I was sitting at the Jazz Festival. See also ALCATRAZ.

Alcatraz

"Alcatraz" is the name of an island near the entrance to San Francisco Bay. The island is, of course, famous for being home to the first ever lighthouse on the Pacific Coast of America. The light was built in 1853 to assist the huge amount of shipping arriving at the port during the Gold Rush.

alcohol

alcohol—usually in the form of methylated spirit—is used on a boat to preheat a paraffin stove (cooker) or lamp; recreational alcohol—usually in the form of beer or liquor—is used on a boat to preheat the crew. (Medicinal alcohol is an antiseptic used for cleaning wounds and abrasions. It works just as well as a stove or lamp preheater. Not beer, of course.)

Overheard in a marina in the Canary Islands as two couples loaded provisions aboard their yacht for the trans-Atlantic passage:

> MAN IN COCKPIT: Did you remember the booze?
>
> MAN ON QUAY: Yes, one case of Scotch, one case of rum, one case of vodka, twenty-four six-packs of beer. . . . Oh, and two bottles of sherry for the women.
>
> MAN IN COCKPIT: Um. I'm not sure where we're going to stow two bottles of sherry.

Alan Villiers said that "Only fools and passengers drink at sea," thus confirming that ocean cruising must be a fool's game. See DEHYDRATION.

a-lee (or a'lee)

1. when the helm is pushed across to the LEEWARD side of the boat, it is said to be "a-lee"; the call "Helm's a-lee!" is often used to tell the crew to release the leeward FORESAIL SHEET and haul in on the WINDWARD side.
2. the side of a boat, island, or some other object away from the direction of the wind.

> The sheets were frozen hard and they cut the naked hand;
> The decks were like a slide, where a seaman scarce could stand;
> The wind was a nor'wester, blowing squally off the sea;
> And cliffs and spouting breakers were the only things a-lee.
>
> They heard the surf a-roaring before the break of day;
> But 'twas only with the peep of light we saw how ill we lay,

We tumbled every hand on deck instanter, with a shout.
And we gave her the maintops'l, and stood by to go about.

All day we tacked and tacked between the South Head and the North;
All day we hauled the frozen sheets, and got no further forth;
All day as cold as charity, in bitter pain and dread,
For very life and nature we tacked from head to head.

Christmas at Sea by Robert Louis Stevenson (1850–1894)

If you want to find out what happened, you will have to read the rest of the poem.

all standing

all the sails are set; also means fully dressed—"He leaped into his BUNK all standing."

almanac

in general terms, an annual publication in the form of a book, booklet, or computer files containing ephemeral information relating to calendar events, anniversaries, and astronomical data; thought to originate from the Arabic *al manakh*.

More specifically, "marine almanacs" and "air almanacs" are published in a joint venture by the US Naval Observatory and HM Nautical Almanac Office. It may come as a surprise that anyone bothers to publish an almanac for air navigators when it must be many decades since an aviator pointed a sextant at the sky in earnest; in the 1920s and 1930s Sir Francis Chichester made many long-distance flights in a Gypsy Moth biplane using astro-navigation techniques (and subsequently all his yachts were named after the airplane). However, since World War II airplanes have been navigated using a combination of DEAD RECKONING, RADIO BEACONS, and GYRO-COMPASSES; and, more recently, GPS.

What keeps the air almanac in print is its popularity with seafaring navigators. The declination and Greenwich Hour Angle of celestial bodies listed in the nautical almanac are accurate to 0.1 minute at intervals of one hour. By comparison, the air almanac is accurate to 1 minute of arc for intervals of 10 minutes—except for the sun and Aries, which are also tabulated to the nearest 0.1 minutes.

aloft

up the mast in the RIGGING making repairs, taking photographs, looking out over a shallow bank of fog, feeling somewhat vulnerable.

amidships

usually refers to the FORE-AND-AFT line of a vessel but more generally might refer to the center section of a boat; when the HELM is "midships," the RUDDER aligns with the fore-and-aft line and, therefore, is encouraging the vessel to travel in that direction, but never guaranteeing that it will.

anchor

a device designed to hook into the seabed and, via a CHAIN or RODE, hold a boat in place.

But it is much more than that, of course. Anchors are a means of security for your beloved (and often uninsured) boat as well as for yourself and your crew. Far more vessels seem to be lost at anchor than on the high seas. Joseph Conrad captured the magic of this lump of metal.

> An anchor is a forged piece of iron, admirably adapted to its end, and in technical language is an instrument wrought into perfection by ages of experience, a flawless thing for its purpose. An anchor of yesterday (because nowadays there are contrivances like mushrooms and things like claws, of no particular expression or shape—just hooks)—an anchor of yesterday is in its way a most efficient instrument. To its perfection its size bears witness, for there is no other appliance so small for the great work it has to do. Look at the anchors hanging from the cat-heads of a big ship! How tiny they are in proportion to the great size of the hull! Were they made of gold they would look like trinkets, like ornamental toys, no bigger in proportion than a jewelled drop in a woman's ear. And yet upon them will depend, more than once, the very life of the ship.
>
> An anchor is forged and fashioned for faithfulness; give it ground that it can bite, and it will hold till the cable parts, and then, whatever may afterwards befall its ship, that anchor is "lost."
>
> The honest, rough piece of iron, so simple in appearance, has more parts than the human body has limbs: the ring, the stock, the crown, the flukes, the palms, the shank.
>
> *The Mirror of the Sea* by Joseph Conrad, 1903

see CONRAD, JOSEPH.

The Black Art of anchors and anchoring should—and does—fill whole books. Here is the short version.

Name	Type	Ground	Comments
Fisherman [kedge, Herreshoff, yachtsman, Nicholson or luke]	Hooking	Rock, coral, clay or hard mud. Weed also; but at your peril.	This is the traditional or "old-fashioned" anchor and more likely to be found in a bar or at the marina gate than in a boat these days. They are seen a lot in developing countries because the patent rights have expired—if they ever existed. The Fisherman lives to haunt us on woolly hats, mugs, and jumpers.
Northill	Hooking	Rock, coral, clay or hard mud.	The Northill looks a little like the Fisherman, but has one marked difference in that the stock is located at the crown, at right-angles to the flukes.
Danforth	Digging	Sand, mud; pulls out more easily than a plough-type.	The Danforth has the singular advantage that it folds flat and is more convenient for stowage. Often used as a kedge and therefore stowed on a rail at the stern. Can pull out fairly easily with sideways force.
Bruce	Plough	Sand, softer mud; the point needs to penetrate.	The flukes of Bruce look like a ploughshare. As the point digs in, the lateral force exerted on the anchor causes (or should cause) it to dig in deeper. The more the pull, the more secure it gets—that's the theory. But the point must dig in for this to work.
CQR	Plough	Sand, softer mud and rocky bottom.	This anchor is very similar to the Bruce, but is hinged at the crown. Its broad flukes mean that it can be effective on rocky ground.

See also ANCHORING AND SEA ANCHOR.

anchorage

any place you might anchor; more usually an area marked on a chart as being suitable. It is a good idea always to check PILOTS or SAILING DIRECTIONS; not all anchorages are safe in all winds. In particular, watch out for desirable anchorages that are also areas in which high voltage power cables have been laid on the bottom. I did this once and got caught—or at least the anchor did, very solidly. When we returned to the same place to get out of a Force 8 wind I had of course learned my lesson; we got snagged up again in the same cable.

anchoring

the procedure of finding a safe place at which to stop, secured by the ANCHOR or anchors with the implication of swimming, food, GUNK-HOLING, EXPLORING ASHORE, and sleep. On the approach, the bow-person is getting the anchor A'COCKBILL and keeping a lookout. At the agreed spot, the anchor is released (or kicked off the roller) and allowed to feed out. Shortly after the chain loses interest it is secured on the GYPSY and the boat is given some reverse to set it. Once the person in the pointed end is convinced it has dug in, more chain can be let out to a length prescribed by the depth. If bad weather is expected this might be a good time to put out a second anchor.

If you are in the Mediterranean and tying up STERN to a pier not far from a Taverna, the above has to be done with some precision to avoid hooking into someone else's RODE.

Such is the importance of good anchoring that I have met many skippers who routinely dive down the chain to ensure that the anchor really is set properly. But not all ANCHORAGES are safe in all conditions and sometimes the best strategy is get the anchor up and confront the weather on the open sea. If the anchor decides it is not going to break free then you may have to drop another, as Jack London explains.

> It is not nice to leave a warm bed and get out of a bad anchorage in a black blowy night, but we arose to the occasion, put in two reefs, and started to heave up. The winch was old, and the strain of the jumping head sea was too much for it. With the winch out of commission, it was impossible to heave up by hand. We knew, because we tried it and slaughtered our hands. Now a sailor hates to lose an anchor. It is a matter of pride. Of course, we could have buoyed ours and slipped it. Instead, however, I gave her still more hawser, veered her, and dropped the second anchor.
>
> There was little sleep after that, for first one and then the other of us would be rolled out of our bunks. The increasing size of the seas told us we were dragging, and when we struck the scoured channel we could tell by the feel of it that our two anchors were fairly skating across. It was a deep channel, the farther edge of it rising steeply like the wall of a canyon, and when our anchors started up that wall they hit in and held.
>
> Yet, when we fetched up, through the darkness we could hear the seas breaking on the solid shore astern, and so near was it that we shortened the skiff's painter.

> Daylight showed us that between the stern of the skiff and destruction was no more than a score of feet. And how it did blow! There were times, in the gusts, when the wind must have approached a velocity of seventy or eighty miles an hour. But the anchors held, and so nobly that our final anxiety was that the for'ard bitts would be jerked clean out of the boat. All day the sloop alternately ducked her nose under and sat down on her stern; and it was not till late afternoon that the storm broke in one last and worst mad gust.
>
> For a full five minutes an absolute dead calm prevailed, and then, with the suddenness of a thunderclap, the wind snorted out of the southwest—a shift of eight points and a boisterous gale. Another night of it was too much for us, and we hove up by hand in a cross head-sea. It was not stiff work. It was heart-breaking. And I know we were both near to crying from the hurt and the exhaustion. And when we did get the first anchor up-and-down we couldn't break it out. Between seas we snubbed her nose down to it, took plenty of turns, and stood clear as she jumped. Almost everything smashed and parted except the anchor-hold. The chocks were jerked out, the rail torn off, and the very covering-board splintered, and still the anchor held. At last, hoisting the reefed main-sail and slacking off a few of the hard-won feet of the chain, we sailed the anchor out. It was nip and tuck, though, and there were times when the boat was knocked down flat. We repeated the manoeuvre with the remaining anchor, and in the gathering darkness fled into the shelter of the river's mouth.
>
> "Small-Boat Sailing" from *The Human Drift*
> by Jack London, 1911

Having been in the situation where we had a Bruce and a CQR out and we were still heading backward toward the beach, take my word for it: choose your anchorage carefully and have an escape strategy should you need it. See BITTS, WINCH, PAINTER.

anchor light

an anchor light (or "riding light") is not just a common-sense safety feature in busy anchorages, it is also a requirement under the COLREGS (US: NAVRULES): Rule 30 states that

> a) A vessel at anchor shall exhibit where it can best be seen:
> (i) in the fore part, an all-round white light or one ball;
> (ii) at or near the stern and at a lower level than the light prescribed in subparagraph (i), an all-round white light.

b) A vessel of less than 50 meters in length may exhibit an all-round white light where it can best be seen instead of the lights prescribed in paragraph (a) of this Rule. (COLREGS72)

It is remarkable the extent to which this rule is flouted, as anyone who has motored into an anchorage at night will attest.

Andrew

a slang expression for the (British) Royal Navy. It has nothing to do with the fact that Prince Andrew was a Royal Navy helicopter pilot and supposedly derives from an 18th-century officer called Andrew Miller. He headed a press gang and, to the best of my knowledge, did not fly helicopters off the back of frigates during the Falklands Conflict.

anemometer

an instrument used for measuring the strength of the wind; modern anemometers fitted to yachts use a digital display to show the strength in miles per hour, kilometers per hour, or meters per second. A graphical display of the APPARENT WIND direction in relation to the yacht is also a useful aid to sailing. If the anemometer is a reasonably sophisticated one and has access to the boat's heading and speed though the water, it will be able to offer you the TRUE WIND speed and direction. (Yes, I know you are supposed to be able to feel the wind on your cheek, but this proves difficult if both cheeks are hidden in the substantial interior of an OILSKIN hood.)

apparent wind

the wind experienced on a boat, being the combination of the TRUE WIND and the movement of the boat through the air; see VECTORS.

archipelago

a large group of islands. The word is thought to derive from the Greek for Aegean Sea, which, of course, contains a large group of islands.

arming

tallow or wax fitted into a hollow in the end of a sounding LEAD, enabling it to bring up a sample of the sea bottom; when such samples (e.g., shell fragments and sand) are matched with a chart, they have the potential to tell you where you are—or, at least, where you are not. The most modern DEPTH FINDER does not have this capability.

ashore

not at sea, not afloat. In the pub maybe?

astern

in the direction of the STERN. A yacht may be traveling backward when "going astern," but not necessarily. The engine may be "going astern" merely as a means of slowing or stopping a yacht's forward motion.

astrolabe

an astronomical and navigation device of Arabic origin dating from about 200 BC, possibly earlier, used for measuring the ELEVATION of CELESTIAL OBJECTS. In its simplest form the astrolabe is a fixed disc calibrated in degrees, with a rotating sighting-bar (an *alidade*) pivoted at its center. In use, the instrument is held vertically while the object is sighted through the alidade; the altitude is then read off the calibrated scale. Over fifteen hundred years the astrolabe became an elaborate and versatile piece of precision engineering incorporating star maps and sundials. For the original user manual for this instrument, see TREATISE ON THE ASTROLABE.

Middle Eastern travelers were, of course, able to use the astrolabe when traveling across deserts (you didn't need a visible horizon) and this may have been the reason why Vladimir Peniakoff took an interest in the instrument. Peniakoff was born in the Ukraine of Russian parents, raised in Belgium, educated in England, and considered himself frightfully British. While working as the manager of a tea processing factory in Egypt in the 1930s he developed an interest in desert exploration. At every opportunity, this fascinating and resourceful man would drive off into the sandy wastes in a modified Austin 7 car. When war reached North Africa, Peniakoff joined the British Army; he was forty-two at the time and had to persuade a doctor to lie about his physical fitness. Because he spoke fluent Arabic and knew the deserts to the west of Egypt he became involved in reconnaissance, ambush, and sabotage behind Italian and German lines. His navigation techniques were based on the use of homemade sun compasses, DEAD RECKONING, and theodolites for star fixes. When he eventually formed his own unit—called Popski's Private Army—he chose the astrolabe as its cap badge.

> I had used as a book-plate for several years a design of an astrolabe, several of my boats had been called by that name, and it seemed now a fitting symbol for a unit that would have to navigate its way by the stars, so I took my book-plate to a Jewish silversmith off the

> Shareh El Manakh and got him to cut in brass a reduced and simplified design of this astrolabe, which we would use as a hat badge. (Astrolabe is a name given to several astronomical instruments formerly used to measure the altitude of stars, before the adoption of the sextant at sea and the theodolite on land. My design was taken from a sixteenth-century Italian instrument.) The first badges were cut and engraved by hand; they turned out rather too exquisite for the roughness of our manners; later we had a die cut and made the badges of silver, which takes the stamp better than brass.
>
> *Popski's Private Army* by Vladimir Peniakoff, 1950

His book relates Popski's adventures in spreading "alarm and despondency" among enemy forces in North Africa and Italy but, sadly, tells us nothing about his escapades on the Mediterranean and the Red Sea.

astrology

ancient belief that the position of the stars and planets influences human behavior and fortune; I have only been able to verify this claim in the case of CELESTIAL NAVIGATION.

astro-navigation

see CELESTIAL NAVIGATION.

athwart

across the boat, at right angles to the CENTER-LINE; from beam to beam, not FORE-AND-AFT.

atoll

a coral reef surrounding a LAGOON. Believed to originate from a Malay word *atolu*, used by Maldive Islanders for their many islands, almost all of which conform to the above definition.

auxiliary engines

engines on sailing yachts are so called because they are the secondary source of power, sail being the first. Many ocean voyagers pride themselves in being able to manage without any auxiliary power. Some sail onto MOORINGS and ANCHORAGES, others row with big oars (Joshua SLOCUM used SCULLING OARS). Sailing into crowded waters however is frowned upon—not least by the owners of adjacent boats. The other disadvantages of being engine-less are the need to rely on other means of generating electricity for light and instruments and the absence of any means of producing hot water in sufficient quantities for a shower.

The choice of engine falls between the two main categories—petrol (gasoline) and DIESEL. Diesel wins out with most boat owners for the following reasons.

- Although diesel engines cost more than petrol engines, they are more than twice as economical to run. A petrol engine consumes 0.454 liters (0.12 US Gallons) per hour per 1 horse-power (0.75 kW). A diesel engine uses 0.227 liters (0.06 US gallons) per hour per 1 horse power (0.75 kW). In most countries, diesel fuel is considerably cheaper than gasoline. The effect of this difference is that diesel-powered yachts need carry much less fuel for an equivalent number of hours under power. And that fuel costs less.
- Unlike petrol engines, diesel units need no electrical power while operating. This is good news on a boat where electrical circuits are usually the first to be attacked by any ingress of water.
- Although electrical starters are convenient for diesel engines, many (especially the smaller ones) can be manually started by using a crank handle.
- For many owners, the choice of diesel power swings on the fact that diesel fuel is much less flammable than gasoline. And, given that fire represents a far greater threat to a boat than the sea, it makes little sense to go sailing with tanks full of the stuff onboard.
- It is the general consensus that, overall, diesel engines are more reliable than gasoline units—another important safety factor.

Q. Who invented the diesel engine?

A. Mr. Herbert Akroyd. See DIESEL ENGINES.

aweigh

an ANCHOR is "aweigh" when it is off the bottom.

awning

a waterproof cover (usually SAILCLOTH) over the DECK or cockpit; more commonly used to keep the sun off in tropical climes; see BIMINI.

azimuth circle/azimuth ring

a metal ring with BEARING SIGHTS fitted around a BINNACLE-mounted compass (usually on the BRIDGE WING of a larger vessel) to enable the taking of compass bearings.

B

back, to

1. a counterclockwise change in the wind direction; e.g., when the wind changes from 180° (South) to 090° (East) it is said to have "backed." See also VEER. It is easy to remember the difference; clocks go *back*—counter-clockwise.
2. backing a sail is a useful technique for maneuvering a yacht; turning across the wind while leaving the sail sheeted on the old LEEWARD side will cause it to back. Doing this with the foresail can be a means of using the wind to push the bow around. If you push the helm to the new leeward side, though, it can cause the boat to HEAVE TO—a quick way of stopping a yacht.

backing down

to maneuver a sport fishing boat in reverse in order to be able to gain line and, eventually, land a fish.

backstaff

a navigation instrument dating from Renaissance times but still in use in the 18th century; while measuring the ALTITUDE of the noon sun, the observer would have his back to the sun. Unlike the CROSS-STAFF, to which it is related, the backstaff is of little use in sighting stars and planets because it functions by casting a shadow from the end of a sliding *upper index* onto a *horizon vane*. The vane includes a narrow horizontal slot; this can be sighted through a pinhole at the end of a sliding *lower index*. The upper index is calibrated for sixty degrees, the lower index for thirty. The observer presets the upper index to a convenient value and then moves the lower index until he can see the edge of the shadow level with the horizon as seen through the slot in the vane. He then adds the setting of the two indexes to obtain the altitude. The backstaff was first described by Captain John Davis in his 1585 book *The Seaman's Secrets* and is, therefore, often referred to as "the Davis quadrant." Its introduction was widely welcomed by navigators for whom blindness in one eye as a result of taking too many noon sun shots had become an occupational hazard.

backstay

a line (usually of heavy wire) running toward the **STERN** from the **HEAD** of a **MAST** to give support to the mast when the wind is from **ABAFT**; some modern yachts have double backstays or a divided backstay—and some have **TENSIONERS** built into them.

backwinding, backing

see **ABACK**

baggywrinkles (bag o' wrinkles)

an important device to prevent chafe on sailing boats. During long periods at sea—especially when running down-wind—the **RIGGING** has a tendency to chafe, the **MAINSAIL** constantly rubbing against the **SHROUDS** and the **SPREADERS** being a particular problem. To counter this, anti-chafe can be made from rope and served around wire and spars when required. Baggywrinkles look like furry sausages of old rope-ends. They look like that because that is what they are. Old manila rope is unlaid (untwisted) and cut to lengths of 8–12cm. The pieces are then woven tightly around a pair of marline-type cords, the first and last being tied off. The resulting baggywrinkles of 30–50cm in length can then be secured to the offending **STAYS** and **CROSSTREES**.

bail, to (sometimes "bale")

to empty water from a boat using any available container; the water is scooped up and chucked over the side. It is said that the fastest way of getting the sea out of a boat is to use a terrified man with a bucket. The word is believed to originate from the Old French *baillier*, to carry, and *baille* for bucket.

ballast

any heavy weight carried in a vessel for the purpose of improving her stability and trim; on yachts, such ballast is carried in the KEEL and serves to counter the HEELING moment of the sails. Ballast can be in the form of water; some trailable yachts have floodable keels that are drained when you need to tow the boat and flooded when you want to sail it. The latest modern ocean racing yachts have ballast tanks in each BILGE and a system of pumps that shifts the weight to the windward side when needed. A cargo vessel is said to be "in ballast" when she has no cargo but is carrying some bulk material in order to maintain good trim.

bar

1. a SHOAL or shallow usually built up at the narrow entrance to an otherwise enclosed waterway; can sometimes produce interesting and entertaining disturbances at certain states of the tide. There is one such free roller-coaster ride at the mouth of Strangford Lough on the east coast of Northern Ireland (or Ulster or the Six Counties depending on your political bent). The tides play interesting games around here. As the tide makes through the narrow entrance, it swirls up surface eddies. Drop the sails and chug into them, put the engine into neutral and release the helm. The eddy will take you through a couple of 360s before getting bored and popping you out the other side.
2. the reason why you are blowing up the inflatable at some odd time of night. Do not consider this to be total madness. It was a common expression in many fishing ports to say that one is "barred" meaning "unable to go to sea because of the state of the passage out of harbor." Now one mostly hears it in a pub: "Do that again and you'll be barred!"

Bar, Crossing the

There has always been a certain mysticism among seafarers about crossing the BAR, whether departing or arriving. This was not lost on Tennyson.

> Sunset and evening star,
> And one clear call for me!
> And may there be no moaning of the bar,
> When I put out to sea,
>
> But such a tide as moving seems asleep,
> Too full for sound and foam,
> When that which drew from out the boundless deep
> Turns again home.

Twilight and evening bell,
And after that the dark!
And may there be no sadness of farewell,
When I embark;

For though from out our bourne of Time and Place
The flood may bear me far,
I hope to see my Pilot face to face
When I have crossed the bar.

Crossing the Bar by Alfred Lord Tennyson

A bit pessimistic, but in Tennyson's era thousands of merchant seamen were lost every year.

Barbary Coast

the North African coast of the Mediterranean Sea. (The name comes from "Berbers" and may also be the root of "barbarous.")

The Barbary Coast is synonymous with the CORSAIRS and PIRACY. Ruthless (and efficient) pirates operated from this coast for 300 years from the 16th century and the whole economy was based on taking hostages for ransom or, alternatively, not taking hostages in return for a "tribute." Many neighboring countries paid millions each year to Berber cities such as Algiers merely in order to stay out of conflict with them.

Berber pirates raided as far as the coast of Ireland and, in 1654, even plundered a Cornish fishing village. Their most famous hostage was Miguel de Cervantes, the Spanish author of *Don Quixote*, whom they lifted in 1575 while he was on his way home from a small war in Italy. Sadly for Miguel he was far from famous at the time of his kidnap so he had to rot away in an Algerian jail for five years before the money was raised to bail him.

Mostly sailing fast LATEEN-rigged ships, the tactics of the Berber Corsairs were well honed.

> Surprise was achieved in a number of ways. A Corsair would hide in the lee of any small island used by merchantmen as a navigational reference point, and then race out and lay alongside in a matter of moments. She might approach down-sun in the evening, making it difficult for the victim to identify her. The Rais [Corsair captain] would hide his fighting men below decks, fly the same colour [*ensign*] as his intended victim, approach within hailing distance, lull the suspicions of the target ship and then suddenly close in. There was always someone aboard the Corsair ship who had as his mother

> tongue the language of any crew likely to sail through the Mediterranean. This man would hail the victim and engage in friendly conversation while the Rais was preparing to pounce. On other occasions prize ships were used as decoys. The Corsair captain relied upon such ruses to claim a quick conquest rather than risk being knocked to pieces by a heavier opponent.
>
> *Gunfire in Barbary* by Roger Perkins and Captain K. J. Douglas-Morris, RN

Many attempts were made to give the Corsairs a good spanking. The British, in concert with the Dutch, bombarded Algiers from the sea in 1816. Almost out of ammunition, Admiral Lord Exmouth conned the Dey into surrendering and freeing hundreds of hostages. But the Barbary Coast survived in business until the mid-1800s.

Most of these pirates were North African or Turkish, but the Brits were to the fore as usual. These included Sir Francis Verney and Captain John Ward.

bare boat charter

a boat chartered (hired) without a crew and available to the charterer to use as he or she wishes, within reason.

barefoot navigation

the concept of back-to-basics, technology-free navigation described in *The Barefoot Navigator* by Jack Lagan (2005).

bare poles

when a sailing vessel has no sails set, it is said to riding under "bare poles" or to be A-HULL.

bargains

shopping around is well worthwhile for many domestic items you need on a yacht. Chandlers tend to be on the expensive side, as do stores with the word "mountain" in their name. Camping stores are worth a try, but so is your local supermarket. It was in such a place that I discovered an excellent range of insulated plastic drinking mugs. They are double-hulled and have a shallow dished removable top with a small hole in it. As you tilt the mug, the tea, coffee or whatever flows into the top and allows you to drink it. The top also helps to keep the drink warm and minimizes spillages. Brilliant! I later saw the same thing in a chandler's for nearly double the price. It did have a little ANCHOR decal on the side, though. Another possible source for more marine-specific bargains are the "boat jumble sales" held in the UK during the spring and summer, as well as car-boot sales, junk shops, charity shops, and even market stalls. In the US I have found shops specializing in second-hand chandlery and they must exist elsewhere.

barge

a large, flat-bottomed wooden or steel boat used for carrying freight on inland or coastal waterways; also used to describe an open ceremonial vessel of the kind on which Cleopatra might have partied. From the Latin *bargia*.

Today, barges are either self-powered or unpowered and towed by TUGS. The earliest rockets used in the American space program were transported to Cape Canaveral in Florida on barges traveling the Intra-Coastal Waterway; a fine confluence of old and new modes of transport.

Perhaps the most successful of the genre was the famed Thames Sailing Barge. Rigged with 456 sq meters (5,000 sq ft) of canvas in a GAFF or SPRITSAIL configuration, these handsome, utilitarian boats could carry 250 tons of cargo. A shallow draft and TABERNACLE-stepped mast made it possible for them to go a long way up-river from London's deep-sea docks. In spite of the absence of a keel, Thames barges are known to have sailed as far south as Spain and as far north as Scotland. When berthed, the SPRIT could double as DERRICKS. The rigs of sailing barges are relatively simple but, as you might imagine, very substantial. In spite of that, barge owners under competitive pressure from steam and diesel remorselessly reduced the crew to a practical minimum of two and even then paid them peanuts.

Although out of business by the 1950s, the Thames Sailing Barge remains a class of boat that rightly attracts enthusiasts determined to keep them afloat. For more on the heyday of these wonderful vessels, the books of barge skipper Bob Roberts are recommended; *Last of the Sailormen* and *Coasting Bargemaster* are published by Sheridan House.

bark

see BARQUE.

barnacle

a marine organism that attaches itself to hard surfaces such as rocks and the hulls of ships.

The goose barnacle (*lepas anatifera*) is the annoying crustacean that fixes itself to the bottom of your boat. As it is taking a ride, the animal in the shell extends limbs called "cirri" that snatch in any passing food. The damn things grow and multiply until they start to slow you down.

The barnacle goose is the smaller European cousin of the Canada goose. It has a white face, black neck, and grey wings but does not stick itself to the bottom of your boat. So what's the connection? For five hundred years from the 12th century, the barnacle goose's means of reproduction was a mystery. John Gerard was an Elizabethan botanist who insisted that he had seen geese breeding from barnacles attached to oak trees by rivers. Rather than ask what he had been smoking, the scientific community believed him. Even Izaak Walton went on about "the barnacles and young goslings bred by the sun's heat and the rotten planks of an old ship, and hatched of trees" (*The Compleat Angler*, 1653). Once it was discovered that the "barnacle" went to the Arctic to breed, common sense prevailed, on this matter at least.

barometer

a device for determining the current atmospheric pressure in millibars; one of the two shiny brass instruments traditionally mounted on the forward BULKHEAD of the main cabin (the other is the CHRONOMETER). There are high-tech versions today that purport to forecast the weather but are best at telling you if it is raining outside. Barometers need to be able to provide an indication of trend; is the pressure rising or falling? Rising = good, falling = bad; it honestly does not get much more sophisticated than that.

barque (or bark)

a sailing vessel with three or more masts, all SQUARE-RIGGED except the MIZZEN, which has a FORE-AND-AFT rig. Three-masted, wooden-hulled vessels of this type predominated as cargo carriers in the mid-19th century but were largely replaced by larger four-masted versions; in America these were called "shipentines." A handful of five-masted barques were built in the 1890s.

barrier reef

a coral REEF that runs parallel to a main shoreline and some distance from it (in comparison with a fringe reef).

batten

a flexible piece of wood or plastic used to stiffen part of a sail; usually slid into pockets sewn into the sail and secured with ties at the LEECH. A Chinese JUNK RIG uses one or more fully battened LUG SAILS where the battens extend all the way from the LUFF to the LEECH.

batten down, to

to secure the HATCHES, PORTHOLES, and any other openings in a vessel in anticipation of heavy weather. This may involve covering with a lashed-down CANVAS or TARPAULIN.

Beach

name used by Marco Polo (1254–1324) for pretty much everything south of China (Cathay), thereby anticipating the vacation resorts of Indonesia and Australia; later known as *Terra Australis Incognita*, or "unknown land down-under." We tend to think of Marco hoofing it but he returned to Venice by sea, departing from Quanzhou and heading south before going through the Strait of Malacca to Sri Lanka, up the west coast of India and overland across what is now Iran. This took him three years (1292–1295) and you could take the same time over the trip today, should you be so inclined.

beachcombing

see EXPLORING.

beacon

a prominent structure on shore (usually near a harbor) established specifically for the purpose of navigation and PILOTAGE.

beakhead, the

the part of the BOW just below the bowsprit (or where the bowsprit would be if you had one); see HEADS.

beam

1. either side of a vessel between the BOW and the STERN; the PORT beam; the STARBOARD beam.
2. the width of a vessel at her widest part; see MEASUREMENTS OF BOATS AND SHIPS.

beam ends

if you BROACH TO and the boat's hull heels through 90°, she is said to be "on her beam ends."

bear away, to (or to bear off)

to change course away from the direction of the wind.

bearing

the direction from one visible object to another (e.g., from a yacht to a LANDMARK); normally measured using a COMPASS (binnacle-mounted or handheld) in DEGREES MAGNETIC, then converted to DEGREES TRUE. In COASTAL NAVIGATION, the intersection of two (preferably three) or more bearing lines on a CHART is an important technique in establishing a position; see COCKED HAT.

beat; beat, to

to sail as close as possible into the wind (usually in a series of zigzag port and starboard TACKS).

Beaufort Scale, the

a scale of wind-speeds used in marine weather forecasting (see table on next page). Originally devised in 1806 by Rear Admiral Sir Francis Beaufort (when he was a mere Commander) it became internationally recognized in 1874 and was upgraded in 1926. Although ostensibly a measure of wind speed, sea conditions tend to be implicit with it.

Up to Force 6 can be considered good sailing conditions (if becoming a little lumpy at Force 6). Anything above Force 6 you do not want to be out in if you can avoid it. Being at sea in Force 10 (Storm) and above has to be considered a survival situation. In a small boat, your life is at risk. Progress is no longer possible and you must engage tactics aimed at saving the vessel and its crew.

becalm, to

when your sailing boat stops because of lack of wind, it is "becalmed." The absence of wind might be caused by entering the LEE of an island or the mainland or by entering an area such as the DOLDRUMS, notorious for long periods of little or no wind punctuated by fierce squalls. In the last case, this might seem like a good time to change the impeller, patch that chafed genoa, try a little fishing. . . . The only problem in the Doldrums tends to be that, although the wind may drop, the sea doesn't and you can end up wallowing uncomfortably for days on end. If the sea eases and you fancy a swim, stayed tied to the boat; it is remarkable how fast conditions can change.

Force	Speed (knots)	Description	Sea State	Wave height (m)
0	0–1	Calm	Like a millpond	0
1	1–3	Light air	Ripples	0
2	4–6	Light breeze	Smooth, small wavelets	0.1
3	7–10	Gentle breeze	Large wavelets	0.4
4	11–16	Moderate breeze	Small waves, fairly frequent white horses	1
5	17–21	Fresh breeze	Moderate waves, many white horses, some spray	2
6	22–27	Strong breeze	Large waves, white foam crests, more spray	3
7	28–33	Near gale	Sea heaps up, white foam streaks from crests	4
8	34–40	Gale	Moderately high waves of greater length, spindrift with foamy streaks	5.5
9	41–47	Severe gale	High waves with tumbling crests, dense foam, and spray affecting visibility	7
10	48–55	Storm	Very high waves with long overhanging crests, heavy tumbling sea	9
11	56–63	Violent storm	Exceptionally high waves, sea covered with long white patches of foam	11
12	64 plus	Hurricane	Air filled with foam and spray, sea white with driving spray, visibility seriously affected	14

becket

an eye or loop fashioned from a rope and often incorporating a THIMBLE.

belay, to

to secure a line (usually an item of RUNNING RIGGING) to a fitting on the boat designated for the purpose (e.g., CLEAT, SAMSON POST, BELAYING PIN).

belaying pin

a short staff of wood (also metal) usually fitted into one of a series of holes through a piece of wood surrounding the foot of a mast or in a plank of wood fitted to the BULWARKS. Positioned as a belay point for RUNNING RIGGING.

bell buoy

a buoy fitted with a fancy bell that sounds as the buoy is moved by the motion of the sea. The bell has four clappers and is mounted in a metal cage. Used to give warning of SHOAL waters in hours of darkness.

belly

the fullness or draught of a sail.

bend, to

to attach two LINES to each other; or a line to one object or between two objects; sails are "bent" to a BOOM or YARD. A "bend" is the name for any knot used to join two lines together.

Bermuda Triangle (The Devil's Triangle, etc.)

an area of sea in which there have been many supposedly mysterious disappearances of ships and aircraft. Many such incidents have actually happened outside the Bermuda-Miami-San Juan triangle, even at the other side of the Atlantic. I have even seen the mystery of the MARY CELESTE (which was found off the Azores) claimed as a mass UFO abduction.

It is certainly the case that the Gulf Stream can be dangerously swift and turbulent and local thunderstorms and water-spouts prevent you getting bored between hurricanes, but as for your thirty-two-footer being beamed up by UFOs or dragged down by giant sea monsters . . . that is all nonsense. Given a choice between alien creatures and bad weather combined with inept seamanship and navigation, you should vote for the latter every time. The US Coast Guard certainly does. The truth of the matter is that there are no more "mysterious" incidents in the Bermuda Triangle than there are in any other similar part of the world's oceans.

Bermudan rig

the standing rig used on most modern yachts; main characteristics are a MAINSAIL and HEADSAIL with no TOPSAIL or BOWSPRIT (and a fore cabin stuffed full of extra sails) . The LUFF of the mainsail is fitted into a groove on the aft side of the mast and the top side of the boom. Variations include the cutter rig, which has an additional foresail hanked to a forestay that leads between the foredeck and a point about two-thirds of the way up the mast, parallel to the main forestay carrying the jib. The Bermudan rig has superior sailing characteristics to windward but many people believe it to be principally a racing configuration (developed in the 20th century) inappropriately adapted to cruising vessels. These people include the author.

This belief, nay conviction, that the last place you want to be on a sailboat in heavy seas and strong winds is the FOREDECK. Not only is it wet and uncomfortable, it is dangerous. So is getting there and getting back to the COCKPIT safely. But on a rig such as this, it is in heavy seas and strong winds that you need to be on the foredeck, forever changing sails to match the wind conditions. The dangers are particularly acute for single-handed sailors who, even if wearing a safety line, will have no one to help them get back onboard. This is the experience of Sir Alec Rose trying to make way down the west coast of Tasmania during his 1967–1968 solo circumnavigation. The date is 18 January 1968 and he is south of Sandy Cape "tired and stiff with the violent motion" from a gale that had just subsided. The sea was calm and it was blowing Force 2 on his bow:

> It was tack and tack to gain a few miles south. The wind went right around the compass twice, and I was kept busy changing sail to try to make the most of it and coax the yacht along. I lowered the mainsail and set the mizzen staysail. I lowered the working jib and set the big genoa. Then the wind went to south and I lowered the mizzen staysail and set the mainsail, lowered the big genoa and set the working jib. So it went on. Whichever tack I went on was the wrong one. By the fifth day I was becalmed off Cape Sorrell, Tasmania, having covered only 330 miles in five days. We ghosted along, getting south.
>
> I got the anchor stowed below deck, and re-stowed some of the stores to lighten the yacht forward. I turned out some old, rusty tins and dumped them overboard.

The elements were to pay Sir Alec back for his decision to finally dispense with the CHICKEN SUPREME. Two days later, the wind started to play tricks again.

> The wind went to the west, very light. I lowered the working jib for boomed-out genoa and lowered the mainsail and set mizzen staysail. The wind went to north-west, and I changed the genoa to the starboard side. The wind increased and with the mizzen staysail the yacht was yawing about unbalanced, so I lowered it. We were getting along well, at six knots, on course to clear southern New Zealand. I set the genoa staysail on the little boom, port side. Then the wind went to south-west, strong, and I had to lower big genoa on starboard side. I hoisted mainsail, stowed genoa staysail and set working jib on boom port side. All this during a few hours, and

> people say, "What do you do with yourself?" I counted up that made sixteen sail changes that day.
>
> *My Lively Lady* by Sir Alec Rose, 1968

Surely, the combined brainpower of those who design, build, and sail yachts can come up with a superior means of managing the means of propulsion than this? One that does not constantly add to the workload of the sailor and place him at risk? Advanced REEFING systems go some way toward solving the problem, but these also add to the other negative features of these rigs—their cost and vulnerability to failure. See also UNSTAYED VS. STAYED RIGS.

berth

1. a fixed bunk on a ship for sleeping in (or recovering).
2. a place in a harbor, marina, or yacht haven where a vessel may be secured.
3. adopted in everyday English as an immeasurable measure of safety: "I would give him/her a wide berth, if I were you."

"between the devil and the deep blue sea"

the "devil" refers to the longest seam to be CAULKED on a wooden hull (fairly high up, in other words); so, if you were "between the devil and the deep blue sea," you were in an intractable position, facing a difficult decision—"between a rock and a hard place."

bight

1. a curve or loop of rope (usually in the middle of the line).
2. a large area of sea between two promontories.

bilge, bilge keels

originally "bulge," the underfloor, the very bottom of the inside of a ship's HULL; a vessel is usually referred to as having two bilges, one each side of the KEEL; some yachts have "twin keels" or "bilge keels"; acts as a repository of all kinds of nasty things from everywhere else on the boat and, therefore, not a nice place to be inverted while changing an IMPELLER. See LIMBER HOLES and PUMP WELL.

bilge pump

as all uncontained water (and other liquids) in a boat will eventually find their way to the BILGE, it is essential that at least one efficient PUMP is set up to drain this area over the side. Yachts can be fitted with a water level detector that automatically starts an electric pump when the water in the bilge reaches a predetermined level. Safety requires that a powerful *manual* pump be installed within reach of the HELMSMAN.

bimini

a sailcloth canopy extending over the cockpit and providing protection from the sun in tropical waters; see CANOPY. Named, one assumes, after the Bahamian island of Bimini.

binnacle

any case or box used to hold a ship's compass on DECK or BRIDGE.

These can be more complicated than they first appear. I once saw a batch of them in production at a delightful Old World company called B. Cooke & Son Ltd. in the English sea port of Hull (Kingston-upon-Hull). The binnacles included location points for corrector magnets to cancel out DEVIATION and a periscope arrangement that enables a compass mounted on a flying bridge to be read below on the bridge proper.

binoculars

an indispensable piece of equipment; the preferred specification is 7 × 50—anything more powerful will make it difficult to hold the object in vision from a moving sailboat because the field of view is too narrow. The "7" refers to the magnification; a buoy seen through the "bins" or "binos" is seven times larger than the buoy seen with the naked eye. The second number refers to the diameter in millimeters of the objective lens (the one furthest from the eye), 50mm in this case. The larger the objective lens, the more light is captured and the more effective will be the binoculars in darker conditions.

Make sure you buy a pair that is waterproof and able to take knocks. If you can afford them, there is a type that incorporates a compass so you can use them to take BEARINGS. See also NIGHT VISION.

biscuits

see HARD TACK.

bitt

a strong vertical wooden post to which a ship's lines are attached. Usually today, on a yacht, a post fitted to the foredeck that (a) provides a heel for the BOWSPRIT (if you have one) and (b) provides a securing-point for the ANCHOR RODE.

bitter end

in general terms, the very end of the line; more specifically, inboard end of an ANCHOR RODE that is tied to a BITT. Now in common English usage: "Our strained relationship had come to the bitter end." See BITT.

bivouac bag ("bivvie bag")

if you have a pilot berth (or QUARTER BERTH) that is susceptible to getting wet, a visit to a camping or army surplus store can provide the solution. Bivouac bags can be cheap PVC or expensive GOR-TEX® and are a waterproof sheath that will completely cover your sleeping bag. Assuming that the berth is stripped down to its plastic-covered mattress, the bivvie will serve to keep your sleeping bag dry whether you are in it or not. (Obviously, make sure that the sleeping bag is tucked well inside whenever hatches are open or the washboards are out.) The "breathable" bivvies are to be preferred; the plastic ones can get very sweaty inside.

Incidentally, bivouac bags sometimes come in the form of a small, one-person tent. These are no use for the above function, but very handy if you want to camp ashore.

blade

the part of a PROPELLER or OAR that does all the work (i.e., the flat bit).

block (block and tackle)

a system of "pulleys" (grooved wheels) mounted in wooden or metal blocks used to increase the mechanical purchase on a line or merely to change its direction. A pair of blocks each with a single central PULLEY or SHEAVE can be arranged in such a way as to improve purchase by a ratio of 2:1; in practical terms, you have to pull twice as much rope through but the arrangement can lift or pull twice the load. By increasing the number of sheaves and the number of runs between the blocks, the purchase ratio can be increased to 4:1, 8:1, or whatever. A traditional low-tech alternative to the use of winches but still found on most boats doing some job or other.

Blue Peter

signal flag used to indicate that a vessel is about to leave port; Flag PAPA in the INTERNATIONAL CODE OF SIGNALS.

bluewater sailing

sailing across an ocean, as against pottering around the bay; the bluer the water, the better.

board, by the

anything lost over the side ("by the freeboard") was the original meaning; now used to mean anything lost or passed over.

boat

see SHIP VS. BOAT.

boathook

a pole with a hook on the end; used for picking up MOORINGS, holding the DINGHY alongside, threatening mutinous crew. . . . Comes in flashy variations: double-hooks, snap-hooks, carbon fiber shafts. Finding somewhere handy to keep the thing on DECK is always a nightmare, so—there is now a Swedish-made collapsible boathook that folds down and fits into a 35cm bag that can be stowed below. This works a bit like the collapsible white sticks used by visually impaired people—so it will be very familiar to many of those who volunteer to head for the PULPIT to pick up the mooring.

Boatswain (Bo'sun)

a senior member of a ship's crew who has responsibility for the maintenance of the vessel. Traditionally, he had other jobs too.

> Captain: The office of the Boatswain is to take into his custody, and to keep under his charge, all the ropes in general belonging to the ship: with all her cables, anchors, and sails; her flags, colours and pendants; and is to stand answerable for them. He is also to take care in peculiar of the long boat and the furniture thereof, and is either himself or his Mate to go in her, and to steer her upon all occasions. He is likewise to call up all the several gangs and companies of men, belonging to the ship, to the keeping of their watches, the exertions of their works and spells (as they call them), and to see that they do them thoroughly; and to keep them in peace, and in order one with another. Lastly, he is (in the nature of a Provost Marshal at land) to see all offenders punctually punished, either at the capstan, or by being put in the bilboes, or with ducking at the main yard-arm; accordingly as they are censured by the Captain, or by a Martial Court.

> Admiral: This officer must needs be of much use and necessity for the due disciplining, and ordering of the whole company belonging to the ship; and it behoves him to be stirring, stout and faithful.
>
> *Six Dialogues about Sea Services* by Nathaniel Boteler, 1685.

It seems odd today that the full job of carpenter, rigger, and sail-maker should be combined with that of onboard policeman.

boatswain's (bo'sun's) chair

a plank of wood with a bridle connected to a rope running up a mast (e.g., the main HALYARD) and used to haul someone to the CROSSTREES or MASTHEAD to carry out repair work; rarely used in this form today on safety grounds. Alternatives include different forms of canvas seats, harnesses, and flexible ladders.

boatswain's (bo'sun's) stores

replacement spares for everything that might go wrong (i.e., break) with the running or standing RIGGING. In the case of ocean passage making, this comes close to carrying a spare set of rigging.

Boat/US

Boat Owners Association of The United States; as well as being the American equivalent of the UK's ROYAL YACHTING ASSOCIATION, Boat/US operates an excellent floating version of an "auto club" rescue service.

bobstay

a chain, rope, or metal rod that runs up from the STEM (bow) of a yacht to the forward end of a BOWSPRIT; used to counter the lifting effect on the bowsprit caused by the FORESAIL. Useful for getting back onboard from a dinghy or after a swim.

body bags

see SEWN UP

bollard

a hefty post of metal or wood solidly built into a quay (at least you *hope* it's solidly built into the quay) and used to secure SHORELINES from a vessel.

bolt rope

a rope stitched into the edge of a sail to both reinforce it and prevent it from fraying; also used to guide the LUFF of a MAINSAIL in a BERMUDA-RIGGED yacht.

bonaventure

name used for a second MIZZEN sail on 16th-century square-riggers; also a good name for a harbor-side bar should you be planning to open one when you SWALLOW THE ANCHOR.

bone in her teeth (or "bone in her mouth")

said of a sailing vessel that is cantering along well enough to throw up a nice white spray or foam under her bow; looks like a bone, right?

> At New York, on those days, the wind howled from the north, with the "storm center somewhere on the Atlantic," so said the wise seamen of the weather bureau, to whom, by the way, the real old salt is indebted, at the present day, for information of approaching storms, sometimes days ahead. The prognostication was correct, as we can testify, for out on the Atlantic our bark could carry only a mere rag of a foresail, somewhat larger than a table-cloth, and with this storm-sail she went flying before the tempest, all those dark days, with a large "bone in her mouth," making great headway, even under the small sail. Mountains of seas swept clean over the bark in their mad race, filling her decks full to the top of the bulwarks, and shaking things generally.
>
> *Voyage of the "Liberdade"* by Capt. Joshua Slocum, 1894

See also CHICKEN SUPREME.

bonnet

an additional panel of canvas or sailcloth laced to the foot of a SAIL to increase its fullness.

books

When he said, "Give me books, fruit, French wine and fine weather," Keats got it right.

boom

a heavy horizontal (FORE-AND-AFT) SPAR used to extend and hold down the foot of a fore-and-aft SAIL (fore, main, or mizzen); see also YARD.

boom crutch

see GALLOWS.

boot stripe

a colored stripe at the BOOT TOP LINE; also known as a "go-faster stripe."

boot top line

the line where the bottom paint meets the topsides paint. This is (or should be) just above the WATERLINE. The boot top line is usually easy to see because the two treatments are often finished in different colors.

bo'sun

see BOATSWAIN.

bottlescrew (US)

a piece of rigging used to connect, usually, SHROUDS to CHAIN PLATES and providing a means of tightening. Central to this device is the turnbuckle that, when turned, draws the shroud and chain plate together.

bottom, the

see SEABED.

bow, bows

the front of a vessel (often referred to as the "pointed end," though some yachts have a pointed end at the STERN, too); "in the bows" merely refers to the front of a ship or boat. "Bow" rhymes with "cow" rather than with "tow."

bower anchor

the main, heavy ANCHOR of a vessel; usually employed with chain at the BOW. See also KEDGE.

bow eye

a sturdy ring fitted to the STEM of a boat and used for towing.

bowline (pron. "bo'-lin")

the very first knot you learn to tie; fashions a safe loop at the end of a rope and is most commonly used to secure SHORELINES to BOLLARDS. It is easiest to learn to tie with the loop *toward* you; sadly though, the first time you tie one in real life, you discover that it's more useful to be able to tie it upside-down from that, with the loop *away* from you. There are variations on the theme: double bowline, running bowline, etc. See KNOTS.

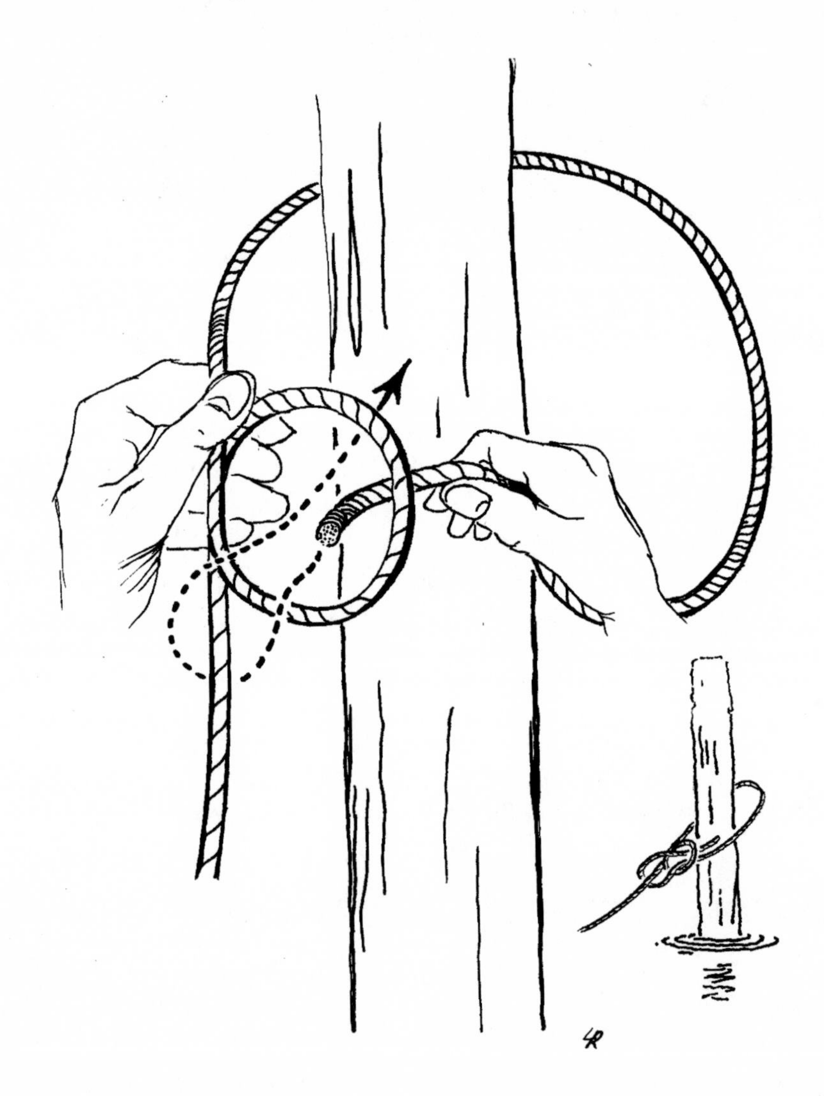

bowser

a truck carrying bulk liquid in an oval or round tubular tank; usually carries fresh water or diesel fuel and should be able to get close enough to your boat for the hose to reach.

bowsprit

a forward-projecting spar at the bow of a yacht to which one or more STAYS are attached. Used to carry FORESAILS with a longer foot (and therefore greater area); most commonly found on traditional boats, but sometimes seen on modern racing yachts such as the J-class. Bowsprits are sometimes retractable in order to facilitate close-quarter maneuvering and reduce berthing fees in marinas.

Although a bowsprit is a "sprit" it does not fall into the conventional definition of "a spar diagonally crossing a sail"; see SPRIT.

boxing the compass

being able to recite the COMPASS POINTS; a way of becoming familiar with a ship's magnetic compass. "Inside fifteen minutes I could box the compass around and back again," boasted Jack London after his first watch at the helm.

brail, a

a line used to gather in a sail and secure it to a mast.

"brass monkey, to freeze the balls off of a"

very cold indeed; supposedly originates from square-rigged warships. The "brass monkey" was a narrow brass tray designed to stop the cannon balls from rolling around the DECK in rough weather. In very cold conditions, the tray would contract, allowing the balls to fall off. It may be time for someone to check the improbable physics of this equally improbable story.

bread

the installation of propane-powered stoves (cookers) onboard cruising yachts (and the availability of dried yeast) makes it possible for bread to be a regular staple item even when offshore. It was not always thus and for centuries sailors had to make do with pretty awful HARD TACK biscuits. The answer to this came from a Mr. Henry Jones.

> A Bristol baker, Henry Jones, had patented a method of making self-raising flour and a bread oven for use on board ship in 1845. He immediately offered his invention to the Navy Victualling Office. After waiting ten years for a positive reply, Jones sent details of his invention and copies of his correspondence with the Admiralty to every Member of Parliament. Within a month self-raising flour was issued to every Navy ship, and for the first time in their history,

> seamen had the pleasure of soft-tack—bread—alongside their ship's biscuit, though it was served only on Sundays.
>
> *The Custom of the Sea* by Neal Hanson, 1999

So, Henry Jones did not just invent self-raising flour, he may also claim to have invented the political lobby industry. See also HARD TACK. Neal Hanson's book is a superbly well-written insight into life at sea in the 19th century.

breakers

waves breaking over rocks, reefs, or shoals; a useful warning sign of trouble to come.

break out, to (break ground, to)

the point at which the FLUKES of an ANCHOR break free of the ground in which they have been holding.

breakwater

a manmade PROMONTORY used to protect a harbor (and the vessels in it) from breaking seas.

breast line

a line that goes directly from the BOW or STERN to the nearest BOLLARD or ring on a quay. See also SPRINGS.

breeze, sea and land

a sea breeze is an ONSHORE wind that occurs in daytime (usually early morning) as the sun heats up the shore relative to the sea (the colder air filling the space made by the rising hot air); a land breeze is the reverse process at night as land cools more rapidly than the sea. The detection of such breezes can provide a navigator with an indication that, although he might not be in sight of land, he or she must be close to it.

Breton Plotter

a navigator's rule incorporating a 360° protractor for the computation of headings; a practical alternative to PARALLEL RULES and the use of a COMPASS ROSE printed on the chart. The Breton Plotter comes into its own on smaller boats where the CHART TABLE is only big enough to accommodate a half- or quarter-folded chart. In this situation it might not be possible to see a compass rose at the same time as the area on which a FIX or course is to be plotted. The plotter carries its own compass rose with it.

bridge

1. an area high on a ship's main superstructure from which the vessel is controlled; the position of the helm.
2. a structure carrying a road, railway, or, rarely, another waterway in the form of a canal. Sailors come across (in the sense of "encounter") bridges more often than you might care to imagine. This is because the banks of river estuaries tend to be populated areas that need to be able to communicate without the inconvenience of a ferry. And the boat's destination might be upstream of the resulting bridge. This presents two interrelated problems: is the underside of the bridge high enough above the surface of the water for the boat—especially yacht—to pass under? And is that still true at high water (because estuaries are invariably tidal)? Charts usually show the clearance height of the bridge at low tide. It is then the job of the navigator to predict (a) an ETA at the bridge and (b) the height of the tide at that arrival time. If (a) minus (b) is greater than the height of your mast plus any antennas you have sticking out of the top, then you will be able to pass through.

The solution to this problem is, of course, a bridge that opens. Probably the finest collection of opening bridges in the world span the Intra-Coastal Waterway on the Atlantic Coast of Florida. There are certain protocols to be followed when going up or down "The Ditch"; hail the bridge master on VHF Channel 9 in good time, ask him or her nicely for an opening and, once the bridge is open, motor through as briskly as you can. Achieving this was, once, a problem for an American friend of mine called Jerry.

> Jerry has to be one of the world's experts on delivering boats on a budget. We met at a riverside boatyard in North Florida.
>
> "What's that guy's name again?" I asked, nodding toward the baby-faced character with spiky blond hair riding into the yard on a bike.
>
> "Jerry," replied the Dutchman. "He's a carpenter working on the old Man O' War Cay sponge boat. Why?"
>
> "His old Man O' War Cay bike's seen better days. . . . So has his haircut. He must get it like that with epoxy. Brakes would be an improvement. On the bike, I mean."
>
> "It's one of those where you just stop pedaling. Doesn't need brakes." A Dutchman would know, of course. And this Dutchman had just solved the problem of how to dispose of a surplus bike.
>
> Hours later, in the local bar, Jerry was proving just how happy he was with his new bikes—he'd helped solve two problems by becom-

ing the proud new owner of a Chinese-built mountain bike *and* a traditional old-style ride-around-Amsterdam-with-your-girlfriend-on-the-handlebars Dutch *fits*—by buying us a beer or three. This also bought him a captive audience and Jerry regaled us with tales of his life as a live-aboard and delivery skipper:

"The only way to make any money on deliveries was to become expert at austerity-budget sailing."

I wondered if he'd ever brewed tea from dinghy rainwater and fried Weetabix for breakfast: "How austere might that be, Jerry?"

"Lunch was a can of beans left on deck for a few hours to heat up. Then eaten outa the can with a spoon."

"That's pretty austere."

He nodded: "Yeah. But once I was taking a boat down the Ditch and ran out of diesel. So, first bridge I came to, I had a problem. I called the bridge master on Channel 9 and told him I had broken down. He opened the bridge straight away and I got the genoa up."

We sipped our Heinekens and waited.

"His timing was perfect and I swept toward the opening bridge on a broad reach. But—darn it—when I fell into the wind shadow the current just carried me back again! I got headway once more and tried again. Meantime, the road traffic was backing up and horns were honking. South Florida was getting close to gridlock."

"So how the hell did you get through?" asked the Dutchman.

"Well," said Jerry, "after three attempts I got the damn boathook and just kinda hauled myself through bouncing off the bridge fenders. Anyway, I decided the situation couldn't go on like that, not with another fifteen bridges to go. I was outa diesel, outa beer, nigh outa Camels and—"

"And right outa luck?" suggested the Dutchman.

"Right outa luck, man! That night I dropped the hook near a bar used by local shrimpers. I rowed ashore with my one remaining asset and explained the situation to them. I rowed back an hour later with a carton of cigarettes, ten gallons of diesel, and a twelve-pack of Budweiser. Next morning I was motoring again."

"Great," we said.

"Everything was going fine until I got to the next bridge."

"Uh-uh," I said.

"Now I couldn't hail the bridge master for an opening—no VHF."

"The one remaining asset?" asked the Dutchman (who knew about assets as well as bikes).

"Yup," nodded Jerry, climbing down from his barstool. "I motored up to the bridge and stood out on the bow to get the attention of the bridge master." He held an imaginary horn to his lips and extended his arm: "Hailing!" Then he touched the fingertips of each hand together and swung his forearms up vertically: "Open bridge!"

"Did it work?"

"Worked just fine," he said, grinning.

The Dutchman met Jerry again a few days later outside the L'il Champ convenience store next to the bar.

"How are you doing, Jerry?"

"Well, kinda OK," he said, tugging at his earring, "but I just spent the night in jail."

"Why? What happened?"

"Well, I was riding that mountain bike you gave me through the Hood." The Hood was right behind the boatyards on the river. "The police stopped me and accused me of stealing the bike. "Nice bike like that," they said, "in a neighborhood like this—you musta stole it!" I told them I damn well *lived* in the Hood and I was on my way home from work and that a Dutchman had given me the bike. They didn't believe me. I said, look officers, if you think *this* is a nice bike, you should see the other one he gave me! It was then they arrested me. My mom bailed me out this morning."

Beware of Dutchmen bearing gifts.

bridge clearance

the distance between the top of your mast (or tuna tower) and the underside of a BRIDGE.

bridge deck

any sort of flat area between the COCKPIT and the COMPANIONWAY; rarely more than that, it seems.

brig

two-masted vessel, square-rigged on both masts. Brigs have been used as both cargo and warships, sporting some ten guns. Length overall was between 25 and 40m (80–130ft) and displacement up to 350 tons. See also BRIGANTINE.

brigantine

a smaller-but-faster two-masted vessel than the BRIG, the important difference being that it was square-rigged on the foremast but the mainmast carried a FORE-AND-AFT MAINSAIL. Some versions carried a square topsail on the mainmast. The name originates from the Italian *brigantino* ("fighting ship") and the Old French *brigandine*. The design dates from the early 1500s. See MARY CELESTE.

brightwork

varnished woodwork or polished metal; the most junior members of the crew get to do the varnishing and polishing.

briney, the

the sea; from "brine," salt water.

bring up, to

to bring a vessel, especially a sailing boat, to a MOORING or ANCHORAGE.

broach; to broach to

the dangerous tendency of a boat running before the wind to swing into the wind thus putting the beam across the waves; can result in CAPSIZING. A sensation not unlike a rear-wheel skid in a car. Usually caused by being over-canvased for the conditions.

buccaneer

a 17th-century seafaring (usually) British businessman dedicated to relieving the Spanish of the unseemly burden of wealth by attacking their ships and colonies on the northern coast of South America. Not to be confused with lawless PIRATES, buccaneers were acting on commission from Her Majesty's government.

The word has interesting origins. "Buccaneer" was originally used to describe someone who made a living from drying meat by slowly heating it on a wooden frame over a fire. (Dried meat was of course the only stuff you could take safely on long voyages.) In the Carib Tupi language, these frames were called *mukem*. Early French settlers rendered this as *boucan* and the folks who used them as *boucaniers*. These *boucaniers* were, it seems, a pretty lawless bunch—hence "buccaneers."

The Haitian version of the *boucan* was the *barbacoa*, from which originates the barbecue. So, the next time you have a cookout, you can impress your guests with this remarkable fact.

bucket-and-chuck-it

the alternative to the dreadful, unreliable HEAD(S); doesn't really need explaining, does it? See also FOA.

bulkhead

a wall inside a yacht or ship; as with vertical partitions dividing the space within a building, those on a vessel are sometimes load-bearing, sometimes not. An important safety feature on a boat is to make it possible to isolate some compartments through the use of watertight bulkheads though this does not work in the case of (1) open doors and hatchways (because they are too fiddly to keep closing behind you), and (2) long holes in the hull that extend across a number of watertight compartments (as in the case of the RMS *Titanic*).

bull rope

see MOORING.

bully beef

corned beef in cans.

bulwarks

a solid wooden "fence" or "rampart" around the DECK of larger ships; stops the worst of heavy seas coming in over the sides and members of the crew going in the opposite direction.

bum boat

those little boats that suddenly appear out of nowhere and surround you as you are dropping the ANCHOR in some distant cove are "bum boats"; goods and services on offer range from water and whiskey, diesel and dog meat, through to a quiet evening with the skipper's sister. In some busy anchorages I notice that the bum boats have gone upmarket and you can call them on VHF and have them deliver, for a price, everything from cold beer, hot pizzas, and replacement IMPELLERS. But I'm not sure about the skipper's sister any more.

bumkin

a spar that projects horizontally from the STERN of a sailboat in the opposite direction from the BOWSPRIT and to which the MIZZEN SHEET is attached.

bunder (*bundha*) boat

a LATEEN-rigged 30m (100ft) LOA general-purpose vessel operating in the Arabian Sea, Persian Gulf, and Red Sea. Much like an Arab DHOW.

When I lived in Karachi (the main port and industrial center of Pakistan) in the early 1970s, bunder boats had a particular appeal. As night fell, a good few of them would make themselves available for charter by young Westerners like myself. The way it worked was this. One would meet a young woman (preferably one who was new to that part of the world—embassy staff were the main target) and invite her out for the evening. After picking her up at her apartment or hotel, one would drive down to the harbor and do a deal with the captain of the most likely-looking bunder boat. You'd get onboard and settle on the comfortable settee in the STERN and the skipper would sail across to a quiet bay in Karachi's huge natural harbor (one of the biggest in the world) and there he would ANCHOR. Under the three-dimensional galaxy show, the crew would bait some lines with small pieces of fish while a charcoal stove was fired up in the bow. A little vibration on the line was the signal to slowly haul in. . . . As the giant crabs came aboard, they were dropped in the pot. The imported white wine had, in the meantime, been chilling over the side. The crab was cracked and served with a sprinkling of fresh lime. . . .

Once a year, the bunder boats were chartered *en bloc* by the yacht club. This annual race was madness itself. You could use the boat's regular crew, but a club member had to helm. Any number of additional crew were allowed—including women and children—but one bottle of Murree beer had to be drunk for every soul onboard. Also, everyone had to wear a silly hat. When I first crewed in this race, in 1971, we won. My main contribution was to slump on a board rigged to windward (to keep the boat upright) and drink as much beer as possible. It was hard, so hard.

bunk

a wooden bed on a yacht or ship, usually arranged FORE-AND-AFT along the inside of the hull and sometimes in tiers; normally fitted with a lee-cloth (sometimes a LEEBOARD or bunkboard) to prevent the occupant from being hurled across the cabin when the helm goes about and suddenly puts his or her bunk on the windward side. A particularly horrible way to be woken up.

bunt

the central part of a sail.

buoy

A floating object varying in shape, color, complexity (and sometimes even position . . .) that is anchored to the seabed; used either for mooring or for navigation purposes. Navigation buoys now fall into distinct classes—LATERAL BUOYS, CARDINAL BUOYS, ISOLATED DANGER MARKS, SAFE WATER MARKS, and SPECIAL MARKS as defined by IALA.

The word probably originates from the Old French *BOIE*, chain. Pronounced differently each side of the Atlantic: as "boy" in English; as "boo-ee" in American.

"We've been around this buoy before"; vernacular expression of the sort used by sales managers meaning, "History repeats itself." (But only because we do not listen.)

burden boards

the floor boards in a wooden dinghy.

burgee

a small triangular-shaped flag flown from the masthead. A burgee may be used for identification (e.g., by a yacht club) or as a simple WIND VANE.

burthen

an obsolete term used to measure a ship's capacity as the number of barrels of wine she could carry; has sadly not been replaced by a term that could describe the capacity of a floating palace in terms of the number of bottles of Gordon's Gin she could carry.

burton, a

a big BLOCK-AND-TACKLE arrangement used to lift a heavy ANCHOR onboard or to tighten the SHROUDS on larger ships.

buying fresh fish

In most tropical ANCHORAGES you are unlikely to be too far from a source of fresh fish. Sometimes, local fishermen will come alongside and offer you a selection of their catch. A bit of local intelligence to get you walking along the WATERLINE in the right direction and at the right time of day could mean you will fetch up at a group of fishermen seine-netting from the beach. And, if you are not too far from a town, there will be a market. Here are some pointers to buying good fish:

- buy where the locals are buying; they will get just as sick as you will from anything that is "off" (have a sniff anyway—it should not smell);
- hold the fish and check that its flesh is firm;
- while you are holding the fish, check that its outer eyes are a shiny, clear golden yellow color and the center black; dull, red and/or sunken eyes are a no-no;
- also look at the scales as the fish bends; they should be shiny and tight—not gaping;
- take a peek under the gill covers—the gills should be red, not brown.

Then it is just a question of agreeing a price. In many places you will be expected to bargain. Do not accept the first price offered, but go for his first compromise and make his day. You can well afford it and he probably has a family at home.

Buys Ballot Law

a simple but crude means of determining the center of the low pressure system influencing the weather you are currently experiencing.

In the Northern Hemisphere, stand with your back to the true wind and the center of the low will be away to your left (low = left) and the corresponding high to your right. In the Southern Hemisphere, stand *facing* the true wind and the center of the low will be away to your left (low = left) and the corresponding high to your right. Credit for this Rule of Arm goes to C. H. D. Buys Ballot, a 19th-century Dutch meteorologist.

"by and large"

sailing with the wind ABAFT the BEAM; just about the only way to make progress in a SQUARE-RIGGER. Now used in everyday English to indicate a generally happy state of affairs.

by the lee

RUNNING downwind with the wind on the same side as the main BOOM; see also LEE, LEEWARD.

cabin

accommodation room in a vessel; will contain one or more BERTHS.

Cabins are rarely perfect because they seem to be designed by people who never have to use them. Tobias Smollet came across a particularly bad example when embarking on a voyage across the English Channel in the late 18th century:

> The hire of a vessel from Dover to Boulogne is precisely the same as from Dover to Calais, five guineas, but this skipper demanded eight, and, as I did not know the fare, I agreed to give him six. We embarked between six and seven in the evening, and found ourselves in a most wretched hovel, on board what is called a Folkstone cutter. The cabin was so small that a dog could hardly turn in it, and the beds put me in mind of the holes described in some catacombs, in which the bodies of the dead were deposited, being thrust in with the feet foremost; there was no getting into them but end-ways, and indeed they seemed so dirty, that nothing but extreme necessity could have obliged me to use them. We sat up all night in a most uncomfortable situation, tossed about by the sea, cold, and cramped and weary, and languishing for want of sleep.
>
> *Travels through France and Italy* by Tobias Smollet, 1766

The descendants of whoever designed this Folkstone cutter are still hard at work today, designing the sleeping arrangements for small boats. And as far as today's ferries are concerned, there are some where a turn of the door handle causes the furniture to be rearranged. See also The CHANNEL (2).

cabin sole

the floor (or floorboards) of the accommodation; keeps your feet out of the BILGES.

cable

1. heavy-duty line (rope, chain, or steel hawser) attached to an ANCHOR or used for towing; see also ANCHOR RODE.
2. a measure of distance at sea, being one-tenth of a NAUTICAL MILE or 200 yards (183m).

cable-laid rope

heavy-duty line made up from three complete ropes twisted together with a left-handed twist.

Camber

1. the curve or slope of a DECK from the CENTER-LINE toward the BULWARKS.
2. the curvature of a sail.

cannabis

see CANVAS

cannibalism

the tradition of eating the most junior member of the crew in a survival situation; sometimes referred to as "the custom of the sea."

Failure to get the PROVISIONING part of sailing right can have gruesome consequences. Over a hundred years ago, when the redoubtable Captain Joshua SLOCUM turned up at the Gilbert Islands (I think it was), he was greeted by

> a canoe coming down the harbour, with three young women in it. One of the fair crew, hailing with the naïve salutation, "Talofa lee" ("Love to you, boss"), asked:
>
> "Schoon come Melike?"

> "Love to you," I answered, and said, "Yes."
> "You come 'lone?"
> Again I answered, "Yes."
> "I don't believe that. You had other mans, and you eat 'em."
>
> *Sailing Alone Around the World* by Capt. Joshua Slocum, 1896

The definitive book on the subject of survival cannibalism is *The Custom of the Sea* by Neal Hanson.

canoe

a narrow, pointed boat, decked or open and constructed from a variety of materials. Canoes are traditionally carved from the trunk of a tree or made by stretching animals skins over a frame (see KAYAK); modern canoes are often made of ALUMINUM or GRP. Propulsion is by means of paddles, but also see SAILING CANOES.

canoe stern

the pointed STERN of a sailing boat that resembles that of a Native American open canoe.

canvas

sail (because it is what they used to be made of); the name of a cloth made from hemp and derived from the Greek word for hemp—*kannabis*. So, the next time customs board you off the Florida Keys and the agent asks if you are carrying any cannabis, you can amuse him with a lecture on the Greek roots of English or you can say, "No, sir!" (The latter is recommended.)

capsize, to

to overturn a vessel.

capstan

a vertically set, concave-waisted rotatable cylinder used for hauling or hoisting CABLE, ANCHOR RODE; may be operated by hand or power; thought to originate from the Latin word *capistrum*. On modern yachts capstans have essentially been replaced by hand-cranked, sometimes power-assisted WINCHES incorporating gearboxes. The originals, some of which stood 1.5m off the deck, were interesting pieces of engineering that enabled the manpower of a ship's crew to take on quite massive loads. Retrieving an anchor was a routine matter but a good winch, a kedge anchor, and a strong crew was often the only means of getting a big vessel off an uncharted shoal and happily floating again. Each essential part of a capstan had its own name: to make the "barrel" rotate the crew would push on "bars" that fitted into square or round holes in the "drumhead"; the vertical ridges on the barrel were called "whelps."

CARD (Collision Avoidance Radar Detector)

a relatively low-cost device that alerts the yacht sailor when he or she is being pinged by someone else's radar. The display indicates the general direction from which the scan is coming. Also incorporates an alarm, thus making CARD pretty essential for the single-hander on passage in or near busy sea lanes and especially in limited visibility. If you have RADAR, a CARD alarm could be the time to switch it on and check for a collision course.

cardboard

the dream material of the packaging industry and just about the worst stuff you can take on a boat. Combined with water it make makes the most unholy soggy mess on the CABIN SOLE, eventually finding its way into the bilges and the sleek black bellies of your live-aboard cockroaches. And if there was food in the cardboard ex-box, then that will be ruined, too. Tip: Don't buy stuff in cardboard. If it only comes in cardboard, transfer it to jars, Tupperware® or self-sealing plastic bags.

cardinal buoys

a system of BUOYAGE used to mark the North, South, East, and West extents of some hazardous obstacle or geographical feature in a seaway; cardinal buoys (named after the cardinal points of the compass) employ a combination of black and yellow colors with double conical TOPMARKS and a system of lights based on the clock (12 or continuous flashing for north, 3 for east, 6 for south, and 9 for west). See IALA.

cardinal points

the four main points of the compass; North, South, East, and West; the half-points are north-east, south-east, south-west, and north-west.

careen

1. careen: a word of Greek origin (meaning "nutshell") used to refer to the HULL or bottom of a boat. It might be based on the resemblance between the seam of a walnut and the protruding keel of a round-bottomed boat. A vessel is "careened" when she is allowed to dry out in such a way that she

settles to one side to allow her bottom to be inspected, scraped, painted or whatever.

2. to careen (US): to go hurtling off at a great rate of knots; also describes the angle of HEEL of a vessel. This form originates in the US in the early 20th century.

carry away, to

when any part of the standing rigging or a spar breaks it is said to have "carried away." What you say is probably unprintable.

Carson, Rachel

The thinking yachtsman's Britney Spears. Most famed for *Silent Spring*, Rachel Carson's books about the sea are *The Sea Around Us* (1950, 1951), *Under the Sea Wind* (1941, 1952), and *The Edge of the Sea*. All are still in print, usually in a combined edition. It is impossible to describe how exquisite her writing is—so here is a sample.

> Eventually man, too, found his way back to the sea. Standing on its shores, he must have looked out on it with wonder and curiosity, compounded with an unconscious recognition of his lineage. He could not physically re-enter the ocean as the seals and whales had done. But over the centuries, with all the skill and ingenuity and reasoning powers of his mind, he has sought to explore and investigate even its most remote parts, so that he might re-enter it mentally and imaginatively.
>
> He built boats to venture out on its surface. Later he found ways to descend to the shallow parts of its floor, carrying with him the air that, as a land mammal long unaccustomed to aquatic life, he needed to breathe. Moving in fascination over the deep sea he could not enter, he found ways to probe its depths, he let down nets to capture its life, he invented mechanical eyes and ears that could recreate for his senses a world long lost, but a world that, in the deepest part of his subconscious mind, he had never wholly forgotten.
>
> And yet he has returned to his mother sea only on her own terms. He cannot control or change the ocean as, in his brief tenancy of the earth, he has subdued and plundered the continents. In the artificial world of his cities and towns, he often forgets the true nature of his planet and the long vistas of its history, in which the existence of the race of men has occupied a mere moment of time. The sense of all these things comes to him most clearly in the course of a long ocean voyage, when he watches day after day the receding rim of the horizon, ridged and furrowed by waves; when at night he becomes

> aware of the earth's rotation as the stars pass overhead; or when, alone in this world of water and sky, he feels the loneliness of his earth in space. And then, as never on land, he knows the truth that his world is a water world, a planet dominated by its covering mantle of ocean, in which the continents are but transient intrusions of land above the surface of the all-encircling sea.
>
> *The Sea Around Us* by Rachel Carson, 1951

No seafarer's library can be complete without the books of this remarkable writer (especially *Under the Sea Wind* and *The Edge of the Sea*).

carvel-built

used to describe the construction of a wooden vessel when the HULL planks are laid flush, edge-to-edge, thus presenting a smooth surface; see CLINKER-BUILT and CAULK.

castaway

the casualty of a shipwreck; as a verb, to be shipwrecked or cast adrift.

Here is a poem by William Cowper to cheer you up during rough weather. It is called *The Castaway*.

Obscurest night involv'd the sky,
Th' Atlantic billows roar'd,
When such a destin'd wretch as I,
Wash'd headlong from on board,
Of friends, of hope, of all bereft,
His floating home for ever left.

No braver chief could Albion boast
Than he with whom he went,
Nor ever ship left Albion's coast,
With warmer wishes sent.
He lov'd them both, but both in vain,
Nor him beheld, nor her again.

Not long beneath the whelming brine,
Expert to swim, he lay;
Nor soon he felt his strength decline,
Or courage die away;
But wag'd with death a lasting strife,
Supported by despair of life.

He shouted: nor his friends had fail'd
To check the vessel's course,

But so the furious blast prevail'd,
 That, pitiless perforce,
They left their outcast mate behind,
 And scudded still before the wind.

Some succour yet they could afford;
 And, such as storms allow,
The cask, the coop, the floated cord,
 Delay'd not to bestow.
But he (they knew) nor ship, nor shore,
 Whate'er they gave, should visit more.

Nor, cruel as it seem'd, could he
 Their haste himself condemn,
Aware that flight, in such a sea,
 Alone could rescue them;
Yet bitter felt it still to die
 Deserted, and his friends so nigh.

He long survives, who lives an hour
 In ocean, self-upheld;
And so long he, with unspent pow'r,
 His destiny repell'd;
And ever, as the minutes flew,
 Entreated help, or cried—Adieu!

At length, his transient respite past,
 His comrades, who before
Had heard his voice in ev'ry blast,
 Could catch the sound no more.
For then, by toil subdued, he drank
 The stifling wave, and then he sank.

No poet wept him: but the page
 Of narrative sincere;
That tells his name, his worth, his age,
 Is wet with Anson's tear.
And tears by bards or heroes shed
 Alike immortalize the dead.

I therefore purpose not, or dream,
 Descanting on his fate,
To give the melancholy theme
 A more enduring date:
But misery still delights to trace
 Its semblance in another's case.

No voice divine the storm allay'd,
  No light propitious shone;
When, snatch'd from all effectual aid,
  We perish'd, each alone:
But I beneath a rougher sea,
  And whelm'd in deeper gulfs than he.

The Castaway (1799) by William Cowper (1731–1800)

"Albion" is an old name for England. See: MAN OVERBOARD; LIFE JACKET; LIFERAFT; JACK-STAYS; SURVIVAL SUITS, and SAFETY LINE.

cast off

to release a rope or cable (e.g., from a CLEAT or BOLLARD).

cat

one of the few animals that is practical to keep onboard a small boat.

Before the innovation of canned food, it was not unusual for livestock to be carried aboard large seagoing ships; pigs, sheep, chickens, even cattle. This was mostly for the benefit of the officers of course, the working crew having to survive on pretty rancid salted beef. See COCKATOO.

catamaran

a twin-hulled boat offering spacious ACCOMMODATION, little stress on the FIDDLES and higher mooring fees. The modern form originates from the Tamil words *kattu* (tie) and *maram* (tree).

I once agreed to crew on the delivery of a 31ft wooden catamaran from North Wales to Oban in Scotland. Late one evening, the owner and his girlfriend were driving us across the north of England to join the yacht. As I catnapped in the back of the car, the owner brought me sharply awake with an unforgettable turn in the conversation.

> "Of course, I built her myself," he said. That little gem hadn't come up in the pub-lunch briefing. But it was to get worse.
> "It's the second one I've built."
> "What did you do with the first one," asked my fellow crew member, "sell it?"
> "No, sadly it sank," said owner, wistfully.
> "At its mooring," muttered girlfriend.

We were traveling too fast for us to effect an escape.

catenary

the deepness of the curve of the cable when riding to ANCHOR.

see ANCHORING.

cathead

metal or heavy timber arrangement projecting from the BOW and used to secure the ANCHOR when raised.

catnapping

a short sleep wherever and whenever possible; sometimes called "power napping." See INFLATABLE PILLOWS.

cat rig

a single-sail rig where the mast is set well forward in the bows; see FREEDOM rig.

cat the anchor, to

to stow the anchor securely after getting under way.

caulk, to

to make a wooden CARVEL-constructed DECK or HULL watertight by hammering (with a caulking iron and a mallet) OAKUM between the planks and covering with hot pitch; hard work for shipwrights but usually paid on a "piecework" basis—by the foot.

cavitation

a phenomenon whereby water is forced away from the working surfaces (blades) of a PROPELLER, thereby reducing its effectiveness and causing premature wear.

cay

pronounced, and in Florida misspelled, "key"; see KEY.

celestial navigation (astro-navigation)

the calculation of a boat's position using a DEAD RECKONING position, an accurate knowledge of GMT, the known position of certain CELESTIAL BODIES (the sun, moon, four planets, and a selection of stars), the ELEVATION of the celestial object obtained using a SEXTANT and a set of mathematical tables to reduce all the above to a reasonably accurate LATITUDE and LONGITUDE. (Within one NAUTICAL MILE is good.) Today, the mathematics may be done using a computer program. A bit slower and a lot less accurate than GPS, but a fine traditional craft that is easier on the batteries. Lunar sight reductions are probably the most difficult.

> As the moon slips from behind a cloud
> And shines,
> So the master comes out from behind his ignorance
> And shines.
> Buddha, from *Dhammapada*, Chapter 13

Sun sights, especially at local noon, are by far the easiest—as Jack London discovered.

> The *Snark* sailed from Fiji on Saturday, June 6, and the next day, Sunday, on the wide ocean, out of sight of land, I proceeded to endeavour to find out my position by a chronometer sight for longitude and by a meridian sight for latitude. The chronometer sight was taken in the morning, when the sun was some 21 degrees above the horizon. I looked in the Nautical Almanac and found that on that very day, June 7, the sun was behind time 1 minute and 26 seconds, and that it was catching up at a rate of 14/67 seconds per hour. The chronometer said that at the precise moment of taking the sun's altitude it was 25 minutes after 8:00 in Greenwich. From this date it would seem a schoolboy's task to correct the Equation of Time. Unfortunately I was not a schoolboy.
>
> *The Cruise of the Snark* by Jack London

Despair not, it comes with practice.

celestial object

an object visible on the CELESTIAL SPHERE that can be used for the purpose of CELESTIAL NAVIGATION; includes some fifty principal stars, four of the planets, the moon, and the sun.

celestial sphere

an imaginary sphere surrounding the earth and sharing the same geographical center and POLES; for the purposes of CELESTIAL (OR ASTRO) NAVIGATION, all celestial objects are assumed to be projected onto the sphere that appears to rotate around the earth in an east-to-west direction. This might seem to be a reckless rejection of everything Polish astronomer Nicolaus Copernicus risked his hide for, but without these assumptions there would be no easy-look-up navigation tables. And that would mean that the computer would have had to be invented before we could have celestial navigation. But it was to produce more accurate navigation tables that the computer was invented in the first place . . . at least that was the purpose of Charles Babbage's mechanical Analytical Engine. It is a good job it turned out the way it did; powering a laptop on an 18th-century square-rigger would have posed a serious challenge. Babbage's "programmer," incidentally, was Lady Ada Lovelace, the daughter the 19th-century romantic poet Lord Byron. Romantic or not, he was pretty sharp when it came to the scientific advancements of the day.

Oh! She was perfect, past all parallel—
 Of any modern female saint's comparison;
So far above the cunning powers of hell,
 Her guardian angel had given up his garrison;
Even her minutest motions went as well
 As those of the best time-piece made by Harrison.

from *Don Juan* by Lord Byron

Byron died in 1824 of marsh fever while trying to organize an uprising in Greece.

centerboard

a board that is lowered through a slot at the center of a sailing DINGHY to reduce sideways movement when BEATING to windward; sometimes also found on smaller cruising yachts designed for shallow waters or for going aground sooner than you need to.

center-line

a notional line running FORE-AND-AFT exactly bisecting a boat.

center of buoyancy (or displacement or flotation)

the average of all forces of buoyancy on a floating vessel can be assumed to function through a single point known as the "center of buoyancy"; the opposite of the center of gravity.

center of effort

the common point through which all motive forces created by the effect of the wind on the sails of a yacht can be assumed to act; a bit like a sideways center of gravity.

chafe, to

to rub, or to damage by rubbing; see BAGGYWRINKLES.

chafe gear

anything used to prevent damage caused by the constant rubbing of one part of the rigging on another; a particular problem when ocean passage making. See BAGGYWRINKLES.

chain locker

stowage space for the ANCHOR chain; because an anchor chain, once retrieved, can have a significant effect on the trim of a sailing boat, it is sometimes fed through a tube to a locker toward the STERN of the forecabin.

chain plate

a metal plate fixed to the DECK or upper HULL of a sailing boat to which the SHROUDS are secured.

Channel

1. the navigable area of a waterway.
2. a radio frequency or, sometimes, a pair of frequencies.
3. "The Channel" refers to the English Channel between Britain and France. Historically, for many Britons, the short voyage across the Channel has been their first journey on a boat. (That includes me; I was fourteen and headed for St. Malo.) In his poem *Dover to Munich* (all right, there were trains involved too), Charles Stuart Calverley described the experience.

> Farewell, farewell! Before our prow
> Leaps in white foam the noisy channel,
> A tourist's cap is on my brow,
> My legs are cased in tourist flannel:
> Around me gasp the invalids—
> The quantity tonight is fearful—
> I take a brace or so of weeds,
> And feel (as yet) extremely cheerful.
> The night wears on:—my thirst I quench
> With one imperial pint of porter;
> Then drop upon a casual bench—
> (The bench is short, but I am shorter)—
> Place 'neath my head the *havre-sac*
> Which I have stowed my little all in,
> And sleep, though moist about the back,
> Serenely in an old tarpaulin.
>
> *Dover to Munich* by Charles Stuart Calverley (1831–1884)

That could have been written last week.

characteristics

see LIGHT CHARACTERISTICS

chart

maps used by seafarers for navigation purposes; contains detail of the coastline, depths, navigation marks such as lighthouses and BUOYS, hazards, and information about tidal currents. Land features are usually restricted to those that are useful for coastal navigation and the whereabouts of pubs and restaurants. Credit for producing the first known chart (1st century AD) goes to a Greek chap called Marinus of Tyre. Marinus decided that the island of Rhodes was at 0° latitude and longitude.

This however did not solve a navigation problem I once encountered while chugging east from Niseros along the Turkish coast toward Rhodes. To pass the time on a windless day, I was taking BEARINGS with a HANDHELD COMPASS, plotting the POSITION LINES and checking the result with a new-fangled handheld GPS set. The resulting COCKED HATS were neat and tight but consistently 2.1nm west of the GPS fix. Dammit, I thought, I'd forgotten to check the chart to make sure the GPS was working to the same DATUM. But it was the same! Maybe it was a DEVIATION problem with the handheld compass?

I had been taking the bearings with my head stuck out of the companionway, so I went up to the BOW and tried again from there. Same problem. Very strange. I dug into the icebox for a beer and sat down to read *all* the small print on the chart. Most charts of the Greek Islands are either British Admiralty charts or editions based on them. This one, however, had been through the hands of the Italian Navy's hydrographers during World War II. And there, in tiny print in the margin, it said, "The accuracy of this chart in some areas cannot be relied upon." Not that British charts were always wonderfully accurate. During the Napoleonic Wars (1803–1815), shipwrecks claimed eight times more lives than cannon shot.

chart plotter

an electric device with a screen used for displaying a chart; usually integrated with GPS.

chart table

a flat surface, rather than a table, on which CHARTS can be laid while the navigator plots positions and courses; usually part of the NAVIGATION STATION so that the navigator can view instruments such as the LOG, wind, DEPTH FINDER, and GPS receiver. The flat surface is usually a flap under which there is stowage space for more charts.

chiccharnies

the little, bearded, and red-eyed tree-dwelling birds unique to the large, sparsely populated island of Andros in the Bahamas; according to locals, these mischievous birds build their nests by weaving the topmost twigs of trees together and are very rarely seen. They are also said to hang from branches by their tails. The original authority on the chiccharnie was the Seminole Native Americans who migrated from South Florida to Red Bay in the north of the island in the 17th century.

Christopher Columbus called Andros *Las Isla del Espirito Santo* (Island of the Holy Spirit), but this mysterious island's modern name is from the British naval commander Sir Edmund Andros. Andros is home to a top-secret NATO research establishment about which less is known than about the chiccharnies.

chicken-fixing, a

see GADGET

chicken supreme

When solo circumnavigator Sir Alec Rose provisioned *Lively Lady* for his 1967 departure from Portsmouth to Melbourne, Australia, the inevitably long list started: "Brown Sugar, 35lb; Tinned Chicken Supreme, 84 tins." . . . Hold

on, now! Not "Tinned Chicken, 84 tins," but "Tinned Chicken *Supreme*, 84 tins." With a little ingenuity and a good selection of herbs and spices you can vary the theme quite nicely on tinned chicken. But chicken supreme will never be anything but chicken supreme. Imagine it . . . you are clear of the English Channel and there is not a single unmanned, auto-helmed container ship in sight and you are *hungry*.

You rig the self-steering for 220°M and within minutes your favorite chicken supreme is simmering on one burner and the rice is bubbling away next to it. You pour an *apéritif*, light a pipe, and relax, appreciating the first really heart-stopping sunset of that summer off the starboard bow, the wake hissing into the distance behind you. You consider the human condition and why you are engaged in the present madness and come to various conclusions, the most significant of which is . . . *that you really hate chicken supreme*! And you have 84 tins of the stuff! It's not even any good as fish bait (to the best of my knowledge).

For a rattling good read, though: *My Lively Lady* by Sir Alec Rose.

chine

the angular edge where the sides meet the bottom of a flat or V-shaped boat.

chock (US)

a FAIRLEAD employed to guide an ANCHOR or mooring line.

chock-a-block

when lines, particularly sheets on an old-rigged ship, are hauled in as tight as possible (as when sailing CLOSE-HAULED) the BLOCKS will tend to get closer and closer together; they will become "chock-a-block." Now used in vernacular English to mean "crowded."

chronometer

a marine chronometer is a clock or watch designed and built for use at sea; chronometers are characterized by their accuracy and reliability. Less important in the days of GPS but were (and still are) essential when CELESTIAL NAVIGATION is used to determine LONGITUDE. Bowditch, the bible of navigation, has a fancy definition:

> The spring-driven **marine chronometer** is a precision timepiece used aboard ship to provide accurate time for celestial observations. A chronometer differs from a spring-driven watch principally in that it contains a variable lever device to maintain even pressure on

> the mainspring, and a special balance designed to compensate for temperature variations.
>
> A spring-driven chronometer is set approximately to Greenwich Mean Time (GMT) and is not reset until the instrument is overhauled and cleaned, usually at three-year intervals. The difference between GMT and chronometer time (C) is carefully determined and applied as a correction to all chronometer readings. This difference, called chronometer error (CE), is fast (F) if chronometer time is later than GMT, and slow (S) if earlier. The amount by which chronometer error changes in 1 day is called chronometer rate. An erratic rate indicates a defective instrument requiring repair.
>
> *The American Practical Navigator* by Nathaniel Bowditch, 2002

The first reliable mechanical marine timepiece, "H4," was invented by John Harrison after it was discovered that a pendulum clock was a waste of space onboard a ship at sea.

> I think I may make bold to say, that there is neither any more mechanical or mathematical thing in the world that is more beautiful or curious in texture than this my watch or timekeeper for the longitude; and I heartily thank almighty God that I have lived so long, as in some measure to complete it.
>
> John Harrison, 1763

Mechanical chronometers are rarely seen onboard ships now and, if you can get your hands on a good one, you may find it will pay for a new auxiliary engine. However . . .

> **Quartz crystal marine chronometers** have replaced spring-driven chronometers aboard many ships because of their greater accuracy. They are maintained on GMT directly from radio time signals. This eliminates chronometer error (CE) and watch error (WE) corrections. Should the second hand be in error by a readable amount, it can be reset electrically.
>
> The basic element for time generation is a quartz crystal oscillator. The quartz crystal is temperature compensated and is hermetically sealed in an evacuated envelope. A calibrated adjustment capability is provided to adjust for the aging of the crystal.
>
> The chronometer is designed to operate for a minimum of 1 year on a single set of batteries. A good marine chronometer has a built-in push button battery test meter. The meter face is marked to indicate when the battery should be replaced. The chronometer continues to operate and keep the correct time for at least 5 minutes while the batteries are changed. The chronometer is designed to

> accommodate the gradual voltage drop during the life of the batteries while maintaining accuracy requirements.
>
> *The American Practical Navigator* by Nathaniel Bowditch, 2002

Will quartz chronometers ever be considered "beautiful and curious"? I doubt it somehow.

Dava Sobel's bestselling little book, *Longitude*, tells the story of John Harrison and is a fascinating insight into the history of the chronometer.

ciguatera

a term used to describe a serious food-poisoning condition caused by eating reef fish carrying an accumulation of *Gambierdiscus toxicus*.

Worldwide, there are over 50,000 reported cases of ciguatera poisoning each year and this is almost certainly an underestimate—it could be as much as one million. It is a condition that seems to have been around for hundreds of years, but only recently has the cause been identified.

Reef fish eat an algae called *Gambierdiscus toxicus*. Then bigger fish such as snapper, grouper, amberjack, and barracuda eat the little reef fish. So far so bad. As the snapper, grouper, amberjack, and barracuda get older and fatter the *Gambierdiscus toxicus* accumulates in their bodies. Then *we* eat the ciguatoxic fish and get very ill. It is impossible to detect ciguatera in a fish by looking at it, feeling it, or smelling it. And it does not matter whether you freeze it, marinate it, or cook it for three days . . . it is still going to get you. You will wake up in the middle of the night with a terrible pain in the belly and feeling nauseated. After spending an hour with your head down the HEAD, you will need to reverse ends; the diarrhea will have started.

If that does not convince you that ciguatera is something to avoid, there is also this:

> Intense itching, joint and muscle pain, tingling of the lips, burning or pain when cold liquids are touched or drank will usually bring at least the more affected victims to the emergency room, later to find most if not all the others who also enjoyed the fish were variably affected. With the exception of ice applied to the tongue or fingertips being interpreted as a burning sensation, there are no clinical signs.
>
> *The Ciguatera Homepage* by Donna G. Blythe, MD

And there is more. These experiences have also been reported: "Symptoms worsened with alcohol, sweets, nuts, or coffee ingestion; headache; double vision and eye irritation; diminished memory; metallic taste; exhaustion; palpitations; tremor; rectal burning/itching; burning sensation while urinating; vaginal pain/tingling; pain in the penis or scrotum; pain with sexual intercourse." Enough already!

The danger zones are any tropical region between the latitudes of 35° N and 35° S. Ciguatera poisoning is endemic in Australia, the Caribbean (all the way from Florida down to the coast of South America), the South Pacific Islands, and Hawaii.

Over 400 species of fish are on the rap sheet but the ones to watch out for are groupers, amberjack, red snappers, eel, sea bass, barracuda, and Spanish mackerel. The size of the fish is an issue and anything much bigger than 2kg can be particularly dangerous. If the whole fish fits on your plate, it will probably be safe (in the sense that it will contain some ciguatera but not enough to strike you down).

claw off, to

to beat to windward to escape a dangerous LEE SHORE; not a situation one should find oneself in the first place—problem avoidance is always preferable to problem solving.

clearing marks

marks used to provide a bearing which, when followed (or kept to PORT or STARBOARD), will ensure a vessel keeps clear of an invisible obstacle.

cleat

a wooden or metal device to which a rope may be secured; usually comprises two upright or curved pieces (or "horns") around which the rope can be bent and tied off; see also JAM CLEAT.

clew

the aft, lower corner of a FORE-AND-AFT sail and the point at or near which the SHEET is fixed.

clews

the ropes that connect the ends of a HAMMOCK to its fixing points.

clinker-built (or clinch-built) (UK)

used to describe the construction of a wooden vessel when the hull planks are laid overlapping; the overlapping parts are called "lands." Copper rivets are used to secure the "lands." The word "clinker" is believed to derive from the verb "to clinch" or "to clench." See LAPSTRAKE (US) and CARVEL-BUILT.

close-hauled

sailing as closely as possible against the direction of the wind; refers to booms of FORE-AND-AFT sails being hauled as close as possible to the CENTER-LINE.

closing

when approaching the line extended from a pair of LEADING or CLEARING marks, the marks will appear to become closer together and are then said to be "closing." See OPENING.

coach roof

a raised area of a yacht's DECK that serves to (a) provide additional headroom in the cabin and (b) facilitate the fitting of windows to provide additional light to the interior.

coaming, coamings

a raised lip around any opening on a DECK (e.g., a yacht's COCKPIT) designed to prevent water from going below.

coastal navigation

navigation carried out within sight of land.

coast guard

an authority or agency (or in the case of the US Coast Guard, an armed service) charged with "policing" a coastline and overseeing the well-being of seafarers.

Coasting

making your way along a coastline.

Also the title of one of the best books ever written about coastal sailing (or indeed about any other kind). It is as much about England and the English and, inevitably, with a book of this kind, about its author, Jonathan Raban, as it is about going to sea. You get a flavor of Raban's adventure from this:

> For four years now, *Gosfield Maid* has been slowly circling round the British Isles. When she first rumbled down the slipway into the Fowey Estuary, I had never taken charge of a boat at sea in my life. A retired naval commander let me play the role of an elderly midshipman, and in a fortnight taught me how to pull up sails, drop anchor, steer a compass course and bleed a diesel engine. In the evenings I taught myself navigation out of books, with the watery yellow lamplight dodging all over the cabin as the boat followed the wakes of passing china clay coasters. On April Fool's Day I left Fowey alone and nervously picked my way out into the English Channel. I had hardly set my course and made my first pencilled cross on the chart before the land faded into the haze.
>
> *Coasting* by Jonathan Raban, 1986.

Not necessarily to be recommended, but the book certainly is.

cocaine

1. an exotic derivative of the coca leaf of South America, especially Colombia and Peru, much in demand in North America and Europe.
2. a white corrosive substance that can (a) cause your boat to be ripped apart even if you don't have any or (b) cause your boat to be confiscated and you thrown in jail if you do.
3. a trading staple of many islands in the Caribbean and SW Atlantic.

> CUT TO: A quiet, care-worn marina somewhere in the Bahamas. The names of people and places have been changed to protect the guilty. So has the name of the MANATEE.
>
> As manatees go, Buzzy wasn't special. He was six feet long and as drop-dead gorgeous as . . . well, as drop-dead gorgeous as any other out-of-place Florida manatee you may have seen. But setting up home in a marina and scoring cabbage leaves from itinerant LIVE-ABOARDS was unheard of around Pot Hole Cay. "Buzzy" soon became the center of attention. He would put in an appearance, wallow around a bit and suckle on the end of a fresh-water hose. Holiday makers in sad shorts would gasp "Gee! Ain't he cute?" and treat him to a vegetarian lunch. However, the sight of a manatee snuffling down the last fresh onion before the arrival of the long-overdue mail boat brought tears to my eyes.
>
> If Buzzy had earned his keep by eating some of the weed growing on the anti-fouling I might have felt differently. I'd been stuck in the run-down marina so long, waiting for parts, that even the NO-SEE-UMS were beginning to amuse me. Then things changed.

One morning, half an hour before dawn, Buzzy and I were rudely woken by a deafening whine. It certainly wasn't the marina's desalinator plant—that had long since stopped working. I turned over and tried to get back to sleep. When the boat started to rock and jerk hard at the shore lines, I was on deck in seconds. Something big and black was hovering a few feet above the mast. My god, I thought, I'm in the BERMUDA TRIANGLE! I looked up and was suddenly transfixed in the beam of an immensely powerful light. . . .

It was the Drug Enforcement Administration, the DEA. I clambered onto the coach roof and shone my big light back at theirs. They retaliated by hovering the CH-46 helicopter even lower and I had to grab the LAZY-JACKS to avoid joining Buzzy in the drink. In the distance, a Coastguard C-130 Hercules circled low and slow. A smaller Blackhawk helicopter made faster passes, its searchlight sweeping the shallow creeks. Now armed Men-in-Black ran through the length of the marina, shining flashlights under the crumbling concrete docks. Maybe they'd come to serve a warrant on the manatee? Maybe Buzzy's tourist visa had expired? But no. Amid all this drama stood an unconcerned woman from a nearby Catalina 42 feeding the aquatic mammal a crisp salad breakfast. What dressing would you like with that, Buzzy? Thousand Island maybe?

M-16 assault rifles at the ready, the Men-in-Black closed in on the SKIFFS tied up by the dockside liquor store and, before long, a voice on VHF Channel 22A announced to the circling Coast Guard plane that "the bodies are in custody." Sure enough, four "bodies" were sitting on the ground, hands cuffed behind their backs, looking glumly up at the troops from the Bahamian Defence Force who had been delivered with such great *son et lumière* by the DEA choppers and the American taxpayer. I recognized the ringleader, the youngest brother of a family whose history went back through rum-running and embargo-busting to piracy. A bespectacled American agent with "Police" emblazoned across the back of his body armor asked the nocturnal traders a few cursory questions before they were rushed to the airstrip in a couple of jeeps. The US Coast Guard Hercules headed off for a hearty eggs-and-steak breakfast on New Providence Island while the DEA choppers made course toward their ultra-secret base at NATO's AUTEC airfield near Fresh Creek on Andros.

Exhibit 1 stayed tied to the dock under the bleary eye of a young Bahamian constable armed with a rusty shotgun older than he was. The suspects' boat was a "600" in island parlance, a cigarette boat with three 200hp Yamahas on the stern and packed beyond the

GUNN'LS with plastic jerry cans of gasoline. Another 600 had been recovered from a nearby beach.

"Why dey have tree outboards?" quipped a local conch fisherman.

"'Coz dey don' have de space for more o' dem, mon," answered the policeman.

No one laughed except me; it was an old joke from the days when, tied alongside every berth here, was a cigarette boat of much confused ownership: "You could buy one for the price of a couple of Kaliks," Evan the barman told us. I was having a lunchtime beer with Dave, the professional skipper from the 85ft "stretch" luxury TRAWLER flying a Cayman Island flag of convenience on the next slip. "They would change hands ten times in as many weeks," Evan continued, "so whenever the Americans turned up with photographs and demanded to know who owned the boats, everyone would shrug. How would we know, mon?"

Dave nodded in agreement: "Yeah, even I had a 600." Evan and I turned and stared at him. "A couple of years ago I woke up and found a cigarette boat on my dock near Lake Worth Inlet in Florida. I reported it to US Customs who found traces of coke all over it." (Dave was talking about the real thing, not the Real Thing.) "Whoever had done the run from over here in the Bahamas had written the boat off as a business expense. The profit had already been made and banked. Customs said that if I made an offer for the boat, I could keep it. One K got me three new 200s."

I was still working out the profit margin by the time I got back to the boat. The Man was waiting on the dock, his legs dangling over the side as he stared at Buzzy. Maybe he was planning to send the manatee back to Florida, a packet strapped under each flipper? The Man knew a thing or two about smuggling.

"Damn stupid, Cap'n!" was his greeting.

"What was damn stupid?" I asked.

"Those guys rehearsing the operation last week—damn stupid." So I wasn't the only one who'd noticed.

The story of Pot Hole Cay is the story of the Bahamas cocaine trade since the 1970s. Just to the west was a cay that no longer seemed to loom large in the day-to-day life of the islands. The stories told about it in the dockside bars were fascinating enough though.

Take the pigs for example. According to Shelley, *Grande Dame* of the marina condominiums, the place was rife with wild porkers of distinctly unfriendly disposition. She told me that a friend once went up there to "walk" his hound. Purely by coincidence, the friend was a retired undercover narcotics agent from Florida. His dog was one of those aggressively friendly breeds—enter a room and he'd be around your neck like a favorite old tie. One day, Shelley's buddy heaved the dog in his whaler and headed west. Once they reached the beach at what we'll call Pig Cay, the dog leaped out and shot off into the brush. Within seconds, it gave a startled yelp and bounded back across the sand, flew into the whaler and cowered behind the spare gas tank.

It could have been a large tusker. From time to time local boatmen would take a day off from diving for conch and go to Pig Cay armed with a 12-gauge shotgun and a few solid slugs and return with the VIP guest for a hog roast.

What might have scared the wits out of Shelley's friend's dog might have been a pig with attitude, but it could have been a snake. Pig Cay has lots of them. There are non-poisonous black snakes. There are also "chicken snakes"—the black and silver fowl snake, related to the boa constrictor. The Dock Master assured me that chicken snakes are harmless. Unless, of course, you are a chicken.

"My son-in-law killed one last week," he told me. "It was nine feet long, even without its head." For fowl snakes still connected to their heads, the problem is this; they wrap themselves around the chicken and hug it to death. Then they swallow it whole, bones and feathers and claws. But instead of crawling off to somewhere discreet, they just lie at the scene of the crime with the chicken in their distended belly while they digest it. This gives the erstwhile owner of the chicken a chance to get both his revenge and a couple of spicy marinated wings.

But the wildlife was not all that was fascinating about Pig Cay. A close examination of the chart shows that it has an airstrip. Shelley, a long-term resident of Pot Hole Cay, told me that she once flew in from Florida with someone she disparagingly described as a "kind of student pilot." As they approached what the apprentice Cessna-jockey thought was the airstrip on Pot Hole Cay, Shelley tapped him on the shoulder.

"You goddamned idiot! Can't you see the oil drums across the runway?!" The plane suddenly lurched into the sky again and the ashen-face pilot circled the airplane for hurried "finals" into Pot Hole Cay International.

A chat with The Man confirmed that Colombian Marching Powder used to be flown on ancient DC-3s into the Pig Cay airstrip from "the south" and then rushed by subcontracted "600s" to Florida's east coast. In those days, everyone made a lot of money. But under pressure from Washington, DC, the Bahamian government disabled Pig Cay airstrip and closed down the island, driving out the "narco-entrepreneurs." When the drug-runners left, the pigs took over the attractive real estate and the snakes moved into the smugglers' abandoned villas and away from the pigs.

"We call dem de Good Ole Days," explained The Man. "Dat's how I paid for de big house."

"And got four years in the other Big House?" I asked.

He grinned: "OK, and four years in jail. But I was just a kid then. Now I'm a respectable businessman." Fair enough; I didn't bother to ask him why he'd needed the waypoints I'd programmed into his Garmin handheld GPS a few days earlier. The last one in the route was a few miles off Miami Beach, the equivalent to Florida's front door. Did this make me an accomplice? God, my fingerprints are on the Garmin!

The downfall of Pig Cay also spelled the end for Pot Hole Cay as a classy resort. By all accounts, the development of the small island was no more than a means of laundering money from the Pig Cay drugs operation. Direct and indirect beneficiaries included professional men and women, Hollywood movie actors, prominent captains of industry, as well as rich sports stars. It would be fun to name names but some of the lawyers were nothing less than media celebrities and I don't fancy spending the rest of my days—and sailing budget—in court. When the drugs profits dried up, so did the laundering and the place fell into decline.

As the sun rose on the day of the big raid on Pot Hole Marina, more light was shed on the pre-dawn drama. The two 600s had been followed all the way from Jamaica by a high-flying DEA Night-Stalker no-see-um. Five huge bales of ganja had been sunk on the flats for later recovery (the cops obliged with this chore the following day), but the cocaine had been dumped into the marina right outside the police inspector's condo.

"Lots and lots of white powder!" his neighbor, Shelley, told me. "It seemed to dissolve very quickly. Does cocaine do that?"

I had no idea. "I've no idea," I told her. But it did occur to me that the high-speed smuggling runs across the Gulf Stream were far from over. I decided the take a straw poll on the matter. There was only one question: Should cocaine be decriminalized? Ninety percent of Americans said "Yes"; 90 percent of Bahamians said "No."

"Legalize it?" laughed The Man. "That would just take all de dollars outa da business, Skipper." That would seem to explain the viewpoint of the Bahamians on the dock—well, at least some of them. I just can't imagine why 90 percent of the Americans I asked voted the other way. . . . Of course, there was the Florida dentist with a big sailboat, a big sport-fishing boat and an even bigger habit. . . . And there was the computer businessman with two sport fishing boats with the same name; he had a notorious nose for the marching powder before he quit and got involved in Florida politics.

It was the highlight of an otherwise boring week waiting for a new propane sniffer and an anchor shackle to arrive from the US. By Saturday, Buzzy was the center of attention again. I was told that he'd been seen racing fast boats to the cut, sticking his head out of the water and clapping his flippers. In return for a rare filet mignon he would even show off his backward somersault.

"I sure wish I could get a pound of those onions," commented a Canadian sitting on the coaming of an aging catamaran.

While picking up provisions the day before our departure for Nassau, I bumped into The Man again.

"Goin' early?" he asked.

"On the tide, such as it is around here."

"You gonna get anudder wake-up call, Cap'n. Tirty-two kees [kilos] comin' in tonight. Dey get busted again."

But the bust no-showed and all was quiet in Pot Hole Cay Marina as we cast off the lines and headed toward the cut, Buzzy keeping close escort off our port bow.

cockatoo

parrots are popular on boats in tropical waters and the cockatoo variety is no exception. Carrie was a fine white crested cockatoo who, each lunchtime in the bar nearest to our mooring, would amuse the clientele with her amazing impersonation of Stevie Wonder. (Think about the head movements.) For her owner, Jake, single and rich in material matters but poverty-stricken in the personality bank, Carrie the Cockatoo was a great way of making friends.

"Aw, my!" the woman at the bar exclaimed, "What a beautiful parrot! What kind is she?"

"She is a very rare Norwegian Blue," I explained, staring into my beer. There was a quiet pause.

"But she ain't really very blue, is she?"

> I shook my head: "No. . . . And she ain't really very Norwegian."
> One day it seemed quiet in the bar.
> "Where's Jake and Carrie?" someone asked.
> "Oh, haven't you heard? Jake's gone back to Boston," explained the bartender.
> "Really? Why?"
> "Carrie the cockatoo died."
> "Aw, no shit! How did that happen?"
> "Well, he was teaching her a new trick and he dropped her and she broke her neck. Jake was real upset. He was showing her how to lie on her back in his hand and stay still. He'd just say to her "Play dead, Carrie!"
> "It worked," came a voice from somewhere near the pool table.

We never saw Jake again.

cocked hat

an expression used to describe the triangle made by the three intersecting position lines drawn on a chart, either from BEARINGS taken of marks on land or of CELESTIAL OBJECTS; the smaller and more regular the cocked hat is, the more accurate the fix.

cockpit

the open part of a yacht in which the crew can sit (or stand) and steer and control the boat using engine and/or sails; most yachts have the cockpit at the STERN, just before the TRANSOM and the MIZZEN mast (if there is one). Central cockpits providing for a cabin in the STERN are also popular and some racing yachts (e.g., the famous Dutch yacht *Flyer*) have a second cockpit just behind the MAINMAST for the FOREDECK crew.

coconut

a European word for a familiar nut. *Coco* is the Portuguese word for a grinning face, most familiar when used for Coco the Clown. Look at the end of a coconut and squint at the three holes. . . . See the face? The name dates from the early 16th century.

> Amongst other things we found here a kind of fruit called *cocos*, which because it is not commonly known with us in *England*, I thought good to make some description of it. The tree beareth no leaves nor branches, but at the very top the fruit groweth in clusters, hard at the top of the stem of the tree, as big every several fruit as a man's head; but having taken off the uttermost

> bark, which you shall find to be very full of strings or sinews, as I may term them, you shall come to a hard shell, which may hold in quantity of liquor a pint commonly, or some a quart, and some less. Within that shell, of the thickness of half-an-inch good, you shall have a kind of hard substance and very white, no less good and sweet than almonds; within that again, a certain clear liquor, which being drunk, you shall not only find it very delicate and sweet, but most comfortable and cordial.
>
> *Sir Francis Drake's Famous Voyage Round the World*
> by Francis Pretty, 1577

While you are holding that coconut, feel the weight. Now imagine it descending at terminal velocity toward your braided skipper's cap from the top of the palm tree 10m above you. Do not fall asleep under palm trees! Each year, eight times more people are killed by coconut strikes than by shark attacks.

cod

a fine white (non-oily) food fish commonly found in the North Atlantic (and once the staple for Britain's fish and chip shops). They used to grow up to 2m long and 115kg, but not anymore. Now sadly an endangered species.

ColRegs (UK, abbr.)

collision regulations; shorthand for International Regulations for Preventing Collisions at Sea 1972, published by the INTERNATIONAL MARITIME ORGANISATION. See NAVRULES (US).

come about

to change HEADING in such a way (and via the bow) that the wind is on the opposite side of the boat.

companionway

stairs leading up and down between DECKS on a vessel.

compass (magnetic)

instrument that indicates direction in relation to magnetic north and enables a vessel to be steered on a COURSE. See also SIDEREAL COMPASS.

compass error

see DEVIATION

compass point

before modern education taught geometry to potential seafarers, the compass was divided into "points." The four cardinal points of North, South, East, and West were each halved to provide NE, SE, SW, and NW—45° slices of the pie. These were then further divided by NNE, ENE, etc. to make 22½° slices. Finally, one further subdivision of the card made an 11¼° slice called simply a "point." See BOXING THE COMPASS, DEGREES.

compass rose

a graphical device resembling a compass card that is printed on a CHART in such a way that it is oriented to the north of that chart. Used to transfer BEARINGS from observations to the chart for navigation purposes.

Conrad, Joseph

master mariner and famed author.

Born of Polish parents in the Ukraine, Conrad fled Imperial Russian persecution as a young man to go to France and become a seaman. His early years found him gun-running in the Caribbean. He became an unsuccessful gambler, sensibly using other people's money. Debts and depression took their emotional toll and he tried to commit suicide by shooting himself in the chest. He failed at that too, managing to miss all vital organs. However, Conrad's 0/10 score for single-handed close-quarter battle was outweighed by his subsequent success as one of the finest writers in the English language. See also SLOCUM, Capt. Joshua.

constellation

a named and recognizable group of FIXED STARS; a centuries-old practice that remains the basis of our ability to locate and identify stars within or in relation to known constellations; the Great Bear (US: Big Dipper) tells us where Polaris (the Pole Star) is.

Constrained Vessel

shorthand for a "vessel constrained by her draft" and defined by COLREGS (US: NAVRULES) International Rule 3 General Definitions as:

> (h) The term "vessel constrained by her draft" means a power-driven vessel which, because of her draft in relation to the available depth and width of navigable water is severely restricted in her ability to deviate from the course she is following.

See RESTRICTED VESSEL.

convection fog

convection fog is, to all intents and purposes, low-level cloud. It differs from ADVECTION FOG in that the lower-level temperature is above rather than below the vapor-laden air.

cookers

see STOVES (COOKERS).

copper-bottomed

descriptive of a wooden ship with a hull covered in copper; in modern-day usage something that is guaranteed, "a copper-bottomed proposal."

Attaching strips of copper to the bottom of wooden ships was an 18th-century innovation intended to make it difficult for barnacles and other destructive encrustations to attach themselves to the hull. The new technique made ships faster and increased the intervals between bottom scraping in dry docks from months to years.

cord

a LINE of one inch (2.5 cm) or less in circumference.

core

center component of a rope or line; covered by the SHEATH.

Corinthian

name for a very wealthy amateur sailor; can also mean something that is ostentatiously luxurious or debauched, or someone from the city of Corinth in Greece. So, in most cases we can ignore the last parameter.

Corsairs

The words PIRATES and Corsairs tend to be used interchangeably, but there is an important difference. That difference is described in a wonderful maritime history book called *Gunfire in Barbary*:

> A Corsair was a professional sailor-raider, the word "corso" meaning chase or pursuit. In principle it is wrong to describe such men as pirates. The act of piracy occurs when robbery is committed at sea, or by attack from the sea, by people who have not been granted any formal authority by a recognised state to commit that act. The

> granting of a letter of marque, signed by a head of state or one of his delegated officers, gave legality to the robbery. Provided that he operated within the limitations set out in his license, the captain could claim that he was practising his profession, not committing a crime. This was an age-old practice, certainly much in favour with the English, many of whose most famous heroes went privateering against the Spanish treasure ships in the Caribbean and South America and much to the benefit of Queen Elizabeth's exchequer.
>
> *Gunfire in Barbary* by Roger Perkins and Captain K. J. Douglas-Morris, RN

This makes the modern world of trade and commerce seem positively unimaginative.

counter

the part of the STERN of a yacht that projects out above the WATERLINE.

counter rotation

two PROPELLERS spinning in opposite directions on the same shaft.

course

generally, the direction in which a vessel is traveling; this might not be the direction in which she is heading. After allowing for the effect of wind and tide, the direction actually traveled is known as "course made good (CMG)" which is strictly speaking a line from the present position (FIX) back to the previous known position. Courses are represented as 0–360° from North (000°), 090° being East, 180° being South, and 270° being West. Allowances also have to be made for MAGNETIC and TRUE headings. The obsolete method of referring to POINTS of the compass—"North," "North-East," "North-by-North-East," etc.—is now only used as a form of verbal shorthand and not at all in the art and craft of coastal and ocean NAVIGATION.

courtesy ensign

the national ensign of a country being visited. This is flown at the starboard CROSSTREE (if you have a crosstree). Anyway, it is smaller than your own ensign, which will be flying at the STERN. It could turn out to be a serious mistake not to fly a courtesy ensign (or to fly the wrong one—like the Turkish ensign when entering Greek waters, for example). Customs and port officials can turn nasty about things like that. Stocking up with the right ensigns for a voyage can be both a planning nightmare and a financial one.

Some sailors take kits of material with them and start sewing up the right ensign at LANDFALL.

cove

a small bay.

coxswain (pron. "coxun")

helmsman, senior member of the crew of a ship's boat. "Swain" is probably a Norse word *sveinn* meaning servant and "cox" derives from the French *coque* for a small boat. In British waters, meeting a coxswain should be a rare but pleasurable experience—he will be helming the RNLI lifeboat.

CQR

a "secure" type of ANCHOR.

crabbing

1. the lateral (sideways) movement of a vessel caused by wind or tidal stream.
2. see BUNDER BOAT.

creek

a narrow coastal waterway.

cringle

a small loop of rope, usually bent and SPLICED back into itself; may include a metal or hard plastic THIMBLE. From the German *kringel* meaning "small ring"

cross-staff

also known as the "fore-staff," the cross-staff was an early, rudimentary instrument used for measuring the ALTITUDE of celestial objects such as the sun, moon, and stars. The device consists of a crossbar that slides along a calibrated staff. The observer holds the end of the staff to his upper cheek and then positions the crossbar so that the top of the bar touches the object at the same time that the bottom is touching the horizon. The altitude is then read off the calibrations on the staff. Being versatile, cheap, and relatively easy to use, the cross-staff was the navigator's instrument of choice for centuries. The downside was that, even with the use of smoked glass shades, the use of this device over a period of time caused blindness.

See also BACKSTAFF.

cross-track error

the difference between a planned position and an actual position; a term used in connection with GPS navigation. In practical terms, a cross-track error (CTE or XTE) is expressed as the distance in NAUTICAL MILES to the left or right of the intended course to the next WAYPOINT or as a new (and constantly changing) bearing to the next waypoint. Navigation by XTE (especially by the pilots of fast powerboats) has become commonplace. This is not particularly good practice if done without reference to the chart because the long curve made by the course over ground might take the vessel over SHOALS or other hazards.

crosstrees (UK), spreaders (US)

most sailing boats today are delivered with the BERMUDAN RIG as standard; and all Bermudan Rigs have on the mainmast—usually about three-quarters of the way up—"spreaders" or "crosstrees." I will call them spreaders for the time being.

Spreaders are commonly shaped like aircraft wings, even to having an airfoil section. They are mostly made from aluminum or carbon fiber and are riveted or bolted to each side of the mast, from which they extend horizontally. "Spreader" is the name used by engineers for anything used to hold cables or wires apart; on a sailing boat they hold the SHROUDS apart. Bermudan-rigged vessels use standing rigging that is constantly under stress and the shrouds that run from deck level to the MASTHEAD are there to keep the mainmast as rigid as possible in the vertical (port-starboard) plane. The greater the angle between the mast and the shrouds, the greater the rigidity achieved. That angle is increased by running the shrouds from the deck to the tip of the spreaders and thence to the masthead. On high-performance boats the spreaders are "swept back" and the CHAIN PLATES securing the foot of the shrouds set aft of the mast, thus providing some resistance to the forward pressure of the sails when running downwind.

"Spreaders" are sometimes called "crosstrees" because they are roughly in the same place on the mast and perform a similar, but not identical function to the crosstrees found on traditional, especially SQUARE-RIGGED ships. When masts were made from the trunks of trees, the height of a mast was, unsurprisingly, restricted by the height of the available trees. But taller masts meant more sails and more sails meant more speed. The solution was to put a small mast on top of the larger, lower one. Where they joined (at "the top") was where crosstrees (made of oak) appeared and they had the job of securing and stressing the upper mast.

crown

the opposite end of an ANCHOR to which the RODE is attached and to which are fixed the FLUKES.

crow's nest

a position high on a mast from which a member of the crew can perform lookout duties; originally improvised from an empty barrel. An important arrangement when hunting whales.

> October 13. "There she blows," was sung out from the mast-head.
> "Where away?" demanded the captain.
> "Three points off the lee bow, sir."
> "Raise up your wheel. Steady!" "Steady, sir."
> "Mast-head ahoy! Do you see that whale now?"
> "Ay ay, sir! A shoal of Sperm Whales! There she blows! There she breaches!"
> "Sing out! sing out every time!"
> "Ay Ay, sir! There she blows! there—there—THAR she blows—bowes—bo-o-os!"
> "How far off?"
> "Two miles and a half."
> "Thunder and lightning! so near! Call all hands."
>
> *Etchings of a Whaling Cruise by* J. Ross Browne, 1846

The crow's nest has now been replaced for all practical purposes by the TUNA TOWER.

cruiser

a boat including comfortable accommodation and designed for cruising rather than racing; see CRUISING.

cruising

the use of a yacht as a means of travel and exploration (unlike yacht RACING, which is a sport). The word "exploration" can of course be taken in its widest, personal sense to mean a search for the inner soul, the meaning of life, or the perfect way of dealing with a surplus of unwanted tinned CHICKEN SUPREME as much as anything of a geological or anthropological nature.

> But do not hurry the journey at all,
> Better that it should last many years;
> Be quite old when you anchor at the island,
> Rich with all you have gained on the way.
>
> *Ithaca*, by Constantinas Cavafy (trans. John Mavrogordato)

Cavafy is, of course, railing against the trend in the second half of the 20th century for the design of yachts to be influenced by the BERMUDAN RIG. Although it can be handled by fewer crew than the traditional LUG RIG, its primary virtue is that it can sail closer to the wind at a higher speed. Yacht designers tell me that a sailing boat can have three characteristics: speed, comfort, and safety—but only two of these can predominate. Who could argue that, for a cruising yacht, comfort and safety must be more important than speed? (And, conversely, are the first to get the chop in the design of ocean racers like the Whitbread 60 class?) In any case, the arguments in favor of the Bermudan Rig are less clear-cut than they may seem at first sight. Modern versions of fully battened LUG RIGS can beat the pants off an equivalent Bermudan Rig on a BROAD REACH or a RUN. And as far as long-distance PASSAGE MAKING is concerned, a smart bit of work with the charts, the pilots, and the weather forecast can easily make up the knot or two difference in windward performance. So, apart from the occasional need to keep out of the wrong sector of a typhoon, what is the hurry? Cavafy was right.

crutch

a U-shaped piece of metal that slots into the sides of a dinghy to serve as the turning point for an oar; see ROWLOCKS.

cuddy (or cuddy cabin)

a small CABIN, usually in the BOW of a small sailing boat such as a DAYBOAT.

Cunningham

arrangement to increase the tension on the LUFF of a sail (usually at a mast); a BLOCK and tackle is used on older yachts, a lever-tensioner on newer ones.

curtains (US)

detachable waterproof enclosures usually fitted each side of the COCKPIT to protect the HELM AREA from weather; see DODGERS.

"cut and run"

to cut your shorelines or anchor cable in order to make a quick getaway.

cutter, a

a single-masted, FORE-AND-AFT rigged yacht with one MAINSAIL and two FORESAILS.

cutwater (noun)

the "cutwater" is the point at which the BOW of a vessel cuts through the water.

cyclone

see TROPICAL STORMS.

D

dagger-board

an adjustable keelboard normally seen on sailing dinghies; raised and lowered vertically by hand and without using a hinge. Ancient seafarers used to thrust lots of dagger-boards down through their "hulls" which is something you can do on a RAFT.

Danforth anchor

an anchor dating from the late 1930s that has folding FLUKES thus enabling it to fold flat on DECK or to fit snugly on the PUSHPIT when available for use as a KEDGE. Tends to skid over weedy bottoms but, once in, it usually stays put. A TRIPPING LINE is recommended. See ANCHORS.

davit(s)

a crane (most commonly with two arms) used for launching and recovering a small BOAT.

dayboat

a small boat without ACCOMMODATION designed for short passages in daylight.

daylight saving time (DST)

an arrangement used in some time zones to vary LOCAL TIME (usually by one hour) during summer to maximize the amount of daylight during working hours; of course, no daylight is actually "saved." You need to be aware of DST because it will be applied to the times shown in local TIDE TABLES.

day sailing

daylight sailing in a small open boat or maybe even a larger one.

daysailor

small, open sailboat raced but sometimes cruised; the supreme example is probably the classic 11ft Mirror Dinghy with its distinctive hard-chine hull. One of these was sailed (and rowed) from North Wales to Romania: *The Unlikely Voyage of Jack de Crow* by A. J. Mackinnon, 2002.

dead ahead

directly ahead, in front of the bow.

deadeyes

a heavy round wooden or metal BLOCK incorporating three holes and used to obtain purchase on rigging such as SHROUDS and RATLINES; the predecessor to the BOTTLESCREW. With a squint, these blocks look a little like skulls, hence "deadeyes." Effectively a six-way block with no sheaves and a lot of friction. Taking up the slack is definitely a three-handed job; one crew member to ease the line, another to pull through and secure, the third to do the cursing. Still used on some boats with quaint rigging, believe me.

Deadhead

1. a hazardous uncrewed floating object; originally a floating log but today more likely to be a nasty low-floating ISO container dropped overboard by a badly and/or over-stacked freighter.
2. a fan of the Grateful Dead rock band.

deadlight

a metal cover that fits over a PORT-LIGHT. Used in bad weather conditions or for additional security when in port and the boat is unattended.

dead reckoning

a means of estimating a position by applying course steered (specifically *assumed* course made good) and distance sailed (speed x time) from a last-known FIX; derived from "deduced ("ded") reckoning" (which is worth remembering because it does not sound nearly as accurate as "dead reckoning"!).

Decca Navigator

an obsolete and now decommissioned radio navigation system.

deck

a platform in a ship covering all or part of the hull area at any level and serving as a floor.

deck boat

a flat open recreational boat usually powered by an outboard engine and seen on inland waters; sometimes called a "party boat."

deckhead

the under surface of the DECK (i.e., the interior surface of the deck, effectively the ceiling of the interior space).

deck light

a small window set into (and flush with) the deck.

deck plate

a metal fitting on the deck that can be unscrewed to gain access to the water or diesel tanks.

declination

a means of describing the position of a star; the "declination" is the angular distance in degrees from the star to the nearest point on the equator of the CELESTIAL SPHERE. "From the star to the nearest point on the equator" of course means "along the GREAT CIRCLE that passes through the star, the celestial poles, and the equator." For the position, needs to be used in combination with the GREENWICH HOUR ANGLE (GHA).

degrees

an arbitrary unit of angular measurement; there are 360 degrees (or 360°) in a complete circle. Degrees are "arbitrary" because logic dictates that they should be metric: 1,000 in a circle would make for easier arithmetic.

dehydration

a potentially fatal condition in which the body lacks fluid; usually caused by vomiting, sweating, and/or diarrhea combined with a failure to take in sufficient water to compensate.

Dehydration is not just a matter of feeling thirsty; in fact, if you are feeling thirsty and drink you are going to be just fine. The primary causes at sea are (1) seasickness, (2) excessive alcohol consumption, and (3) "stomach bugs" or more seriously dysentery. In the case of seasickness, victims must be encouraged to drink non-alcoholic liquids as frequently as possible—especially if most of it is still being vomited over the side. In the case of alcohol, a few beers or a "tot" need not be a serious problem. Hangovers are caused by dehydration and dehydration is caused by too much alcohol in the system. Heavy drinking at sea is not a wonderful idea and skippers should take action if they are concerned about the speed at which liquor stocks are being consumed. An immediate measure is to persuade the "patient" to drink a lot of water *before* turning in for the night.

If none of this seems to be working, explain in detail to the patient that the next step will be to administer water by way of an enema. That usually gets everyone back on their feet again.

dengue fever

see MOSQUITO.

depth finder

an instrument used for determining the depth of the sea under a vessel; specifically, the distance between the position of the through-hull transducer (which certainly will not be at the bottom of the keel) and the sea bed. This discrepancy can be entered into most modern depth finders so that the instrument always displays the depth from the bottom of the keel.

Something had to replace swinging the LEAD and this is it. Depth finders work by sending an acoustic "beep" from a transducer fitted through the hull toward the bottom of the sea beneath the boat. The signal bounces back and the time it takes to do this is measured and translated into a display showing feet, FATHOMS, or meters. In the earliest versions of these devices the display used a circular scale running from 0 to (say) 100 fathoms on which the depth showed as a flickering LED bar of light. A little more upmarket and more useful was a pen on the end of an armature that transcribed the depth as a continuous line on a moving scroll of paper. This enabled the navigator to see a cross-section of the bottom over which he was sailing.

Low-cost modern depth finders use a digital display; a push of a button switches the units between feet, fathoms, and meters. For a few pennies more, you can have a display showing a running profile of the bottom. The latest version of these "scans ahead" so you can judge the depth in advance—a useful aid to avoid going aground. Even higher-resolution models are good enough to show shoals of fish—or even larger individual marine animals. These are more commonly called "fish-finders" and the next generation will be able to tell you which bait to put on the hook. . . .

depth sounder

see DEPTH FINDER.

derrick

a hinged spar, to the outer end of which is fixed a TACKLE that can be used for lifting heavy objects. Originally employed for handling cargo, now more likely to be used on a yacht for launching and recovering a TENDER and be called a DAVIT or davits. A good rig to have—even if improvised—because it can be used to help recover anyone who might have fallen in.

The word originates from the name—"Derick"—of a famous 16th-century Tyburn hangman who was, it seems, a master with the gallows. Tyburn was at Marble Arch in London.

destroyer

a fast lightly armored warship heavily armed for a defensive role against aircraft and submarines.

deviation

an inaccuracy in a COMPASS reading caused by magnetic influences from surrounding metal on the vessel in which the compass is mounted (or maybe by that portable radio you wedged next to the BINNACLE?); not to be confused with VARIATION, which is a natural phenomenon. It seems to be a common

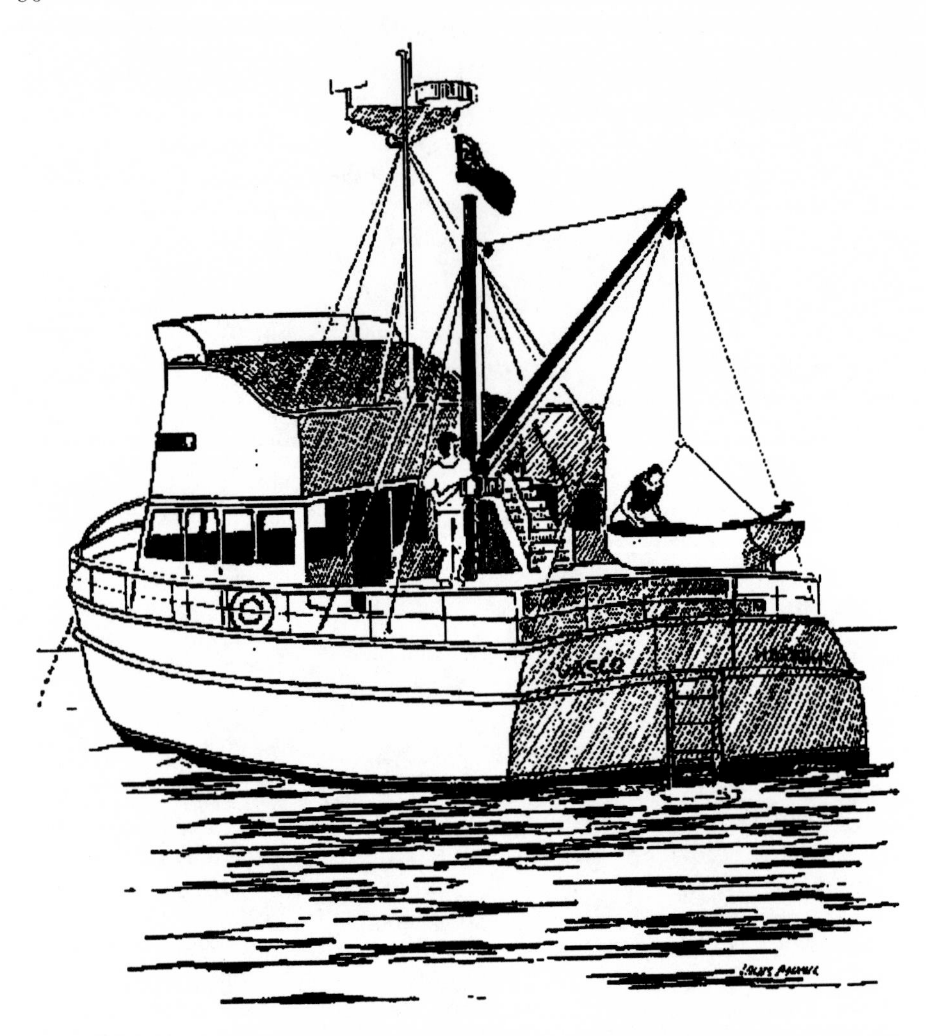

fallacy that deviation does not apply to hand-bearing compasses, so try taking the same bearing from different parts of your boat (keeping well clear of the cockpit stereo speakers, especially if you are into Metallica or any other form of heavy metal).

Devil's Triangle, The

see BERMUDA TRIANGLE.

diesel engines

an internal combustion engine that operates on the compression-ignition principle as follows:

- the first downward stroke of the cylinder draws in air only;
- the compression stroke causes the air to heat to a very high temperature (the Law of Thermodynamics applies);
- at the top of the compression stroke, an "injector" squirts in a small amount of vaporized fuel that ignites;
- the resulting explosion forces the cylinder down, giving it energy that is transferred to the crankshaft and, via the gearbox, to the prop shaft;
- the cylinder then rises again to blow out the gases and start the cycle again.

The origins of the "diesel" engine lie in the latter half of the 19th century when the race was on to come up with a cleaner, more efficient replacement for the conventional steam engine. The principle of the diesel was first drawn up in 1890 by an Englishman with the delightful name of Herbert Akroyd. But it was not until three years later that a German engineer called Rudolf Christian Karl Diesel actually patented the diesel engine as we now know it.

Rudolf's idea was to invent an engine that would provide limitless cheap power in developing countries. Early versions used coal dust as fuel; then there were experiments with vegetable oils before a refined mineral oil byproduct was selected. The power of the in-cylinder explosions is such that the casings of a diesel engine have to be very strong. During one early test a cylinder head blew off, nearly taking Rudolf's head with it.

Born in Paris in 1858, Diesel died fifty-five years later in mysterious circumstances, disappearing over the side of a ferry. I assume that Herbert Akroyd was not a suspect.

In recent years, in UK waters, the most common cause of distress calls to the RNLI lifeboat service has been engine failure (and I suspect the same is true worldwide). In response, the ROYAL YACHTING ASSOCIATION introduced a one-day course on diesel engine basics and how to avoid or deal with the usual causes of failure (water or air in the fuel supply, failure of the impeller used to pump sea water around the cooling system). This is a good course that probably has equivalents in other countries and is a recommended investment of time and money for all seafarers.

Differential GPS

an ingenious system for countering the effect of SELECTIVE AVAILABILITY in the GLOBAL POSITIONING SYSTEM.

The basic principle is this. A station at a fixed, known position receives the current GPS signal and deduces a position. The difference between the known position and the GPS position is called the "differential." This differential is then broadcast (usually on local FM frequencies). Vessels with DGPS receivers can then apply the differential to their own GPS position, thus improving its accuracy from about 100m to less than 1m, maybe even 30cm. Service normally provided in busy port areas and operated by either an appropriate local authority or by a commercial company for a usage fee extra to the cost of the DGPS receiver. For large commercial vessels such accuracy enables the skipper to know not just the position of the vessel, but the exact position of its BOW and STERN. DGPS is probably unnecessary for recreational boaters and yachtsmen. See GLOBAL POSITIONING SYSTEM, DYNAMIC POSITIONING.

dinette

the correct name for that area of the main cabin with a table and upholstered bench seats on two or three sides; on smaller boats might double as a navigation table and, at night, convert into a double berth.

dinghy

an open dayboat, which can be rigged for rowing, sailing, or motoring (with an outboard); usually constructed from wood, GRP, or inflatable synthetic rubbers. In the last case the dinghy is often called an "inflatable." Commonly used on yachts as a TENDER.

When going for a sail on someone else's boat, one's first encounter is likely to be with the dinghy. When the spy Carruthers met up with Davies in Flensburg he was expecting a paid crew. But Davies had been single-handing the *Dulcibella*. And the boat was at anchor.

> "If you'll get into the dinghy," [Davies] said, all briskness now, "I'll pass the things down." I descended gingerly, holding as a guide a sodden painter which ended in a small boat, and conscious that I was collecting slime on cuffs and trousers.
>
> "Hold up!" shouted Davies, cheerfully, as I sat down suddenly near the bottom, with one foot in the water. I climbed wretchedly into the dinghy and awaited events. "Now float her up close under the quay wall, and make fast to the ring down there," came down from above, followed by the slack of the sodden painter, which knocked my cap off as it fell. "All fast? Any knot'll do," I heard, as I grappled with this loathsome task, and then a big, dark object loomed overhead and was lowered into the dinghy. It was my portmanteau, and, placed athwart, exactly filled all the space amidships.
>
> "Does it fit?" was the anxious inquiry from aloft.
>
> "Beautifully."
>
> "Capital!"

Scratching at the greasy wall to keep the dinghy close to it, I received in succession our stores, and stowed the cargo as best I could, while the dinghy sank lower and lower in the water, and its precarious superstructure grew higher.

"Catch!" was the final direction from above, and a damp soft parcel hit me in the chest. "Be careful of that, it's meat. Now back to the stairs!" I painfully acquiesced, and Davies appeared.

"It's a bit of a load, and she's rather deep; but I *think* we shall manage," he reflected. "You sit right aft, and I'll row."

I was too far gone for curiosity as to how this monstrous pyramid was to be rowed, or even for surmises as to its foundering by the way. I crawled to my appointed seat, and Davies extricated the buried sculls by a series of tugs, which shook the whole structure, and made us roll alarmingly. How he stowed himself into rowing posture I have not the least idea, but eventually we were moving sluggishly out into the open water, his head just visible in the bows. We had started from what appeared to be the head of a narrow loch, and were leaving behind us the lights of a big town. A long frontage of lamp-lit quays was on our left, with here and there the vague hull of a steamer alongside. We passed the last of the lights and came out into a broader stretch of water, when a light breeze was blowing and dark hills could be seen on either shore.

"I'm lying a little way down the fiord, you see," said Davies. "I hate to be too near a town, and I found a carpenter handy here—There she is! I wonder how you'll like her!"

I roused myself. We were entering a little cove encircled by trees, and approaching a light which flickered in the rigging of a small vessel, whose outline gradually defined itself.

"Keep her off," said Davies, as we drew alongside.

The Riddle of the Sands (1903) "edited"
by Erskine Childers (1870–1922)

The only experience that really rivals the fairly commonplace encounter described by Childers is when you are the first to arrive, there are a dozen dinghies tied up to the pier, and the owner has provided only the vaguest indication of where the boat is ANCHORed. Oh, and the batteries in your flashlight are flat.

dip

1. a correction expressed as a number of degrees made to sextant readings to allow for the navigator's sextant being some height above the horizon; varies according to the "height of eye" above the surface: the greater the height, the bigger the dip.

2. to dip is to lower and re-hoist an ENSIGN as a salute to another vessel that should respond by dipping its own ensign. Try it the next time you sail past an aircraft carrier!

direct drive

a transmission configuration in which the drive shaft runs in a straight line from the inboard engine, through the gearbox, and through the hull to the propeller.

direction of buoyage

the direction of passage a vessel is assumed to have when describing a system of BUOYAGE; IALA System A assumes that a vessel is entering a seaway, System B that it is leaving. The concept is also important in the case of channels that have two or more entry points from the open sea. Every chart covering such an area (e.g., the English Channel, the Menai Straits between Wales and Anglesey) bears symbols (big arrows) indicating the prevailing direction of buoyage.

displacement

the weight of a boat or ship measured in terms of the weight of water the vessel displaces; see MEASUREMENTS OF A BOAT OR SHIP.

displacement hull

a boat that moves through the water rather than over it (as in the case of a planing boat).

Distress Signals

the approved distress signals for use at sea are specified in Annex IV of the COLREGS (US: NAVRULES). It is worth making yourself well familiar with all of these options.

DISTRESS SIGNALS

1. Need of assistance

The following signals, used or exhibited either together or separately, indicate distress and need of assistance:

- a) a gun or other explosive signal fired at intervals of about a minute;
- b) a continuous sounding with any fog-signaling apparatus;
- c) rockets or shells, throwing red stars fired one at a time at short intervals;
- d) a signal made by radiotelegraphy or by any other signaling method consisting of the group . . . - - - . . . (SOS) in the Morse Code;

e) a signal sent by radiotelephony consisting of the spoken word "Mayday";
f) the International Code Signal of distress indicated by NC;
g) a signal consisting of a square flag having above or below it a ball or anything resembling a ball;
h) flames on the vessel (as from a burning tar barrel, oil barrel, etc.);
i) a rocket parachute flare or a hand flare showing a red light;
j) a smoke signal giving off orange-coloured smoke;
k) slowly and repeatedly raising and lowering arms outstretched to each side;
l) the radiotelegraph alarm signal;
m) the radiotelephone alarm signal;
n) signals transmitted by emergency position indicating radio beacons;
o) approved signals transmitted by radiocommunication systems;

2. The use or exhibition of any of the foregoing signals except for the purpose of indicating distress and need of assistance and the use of other signals which may be confused with any of the above signals is prohibited.
3. Attention is drawn to the relevant sections of the International Code of Signals, the Merchant Ship Search and Rescue Manual and the following signals:
   a) a piece of orange-coloured canvas with either a black square and circle or other appropriate symbol (for identification from the air);
   b) a dye marker;

From COLREGS72. Option 1(h) may, of course, be the reason for your distress.

doghouse

a small (usually square) WHEELHOUSE built around the COMPANIONWAY and providing protection from the weather.

Doldrums, the

low-pressure areas around the equator in the Atlantic Ocean and the Pacific Ocean where the prevailing winds are calm but unpredictable.

> Wind is caused by the need for air to move from one place to another, specifically from a place of high pressure to one of low pressure. The tropical zone is the part of the Earth squarest-on to the Sun so it tends to be hotter than anywhere else. As the hot air expands it rises, causing colder air from the north and south to rush in and fill the gap. The so-called "sub-tropical high pressure system" results in north-easterly trade winds in the northern hemisphere and south-easterly trades in the southern hemisphere. Where they meet is called the "Inter-tropical Convergence Zone" (ITCZ). The influence of the ITCZ extends as far as 30 degrees north and 30 degrees south. Sandwiched between these is an area known as the "Doldrums" where the winds are light, variable or non-existent and punctuated by fierce squalls.
>
> from *The Barefoot Navigator* by Jack Lagan

The Doldrums are hated by sailors. Samuel Taylor Coleridge explained why in *The Rime of the Ancient Mariner*:

> All in a hot and copper sky,
> The bloody Sun, at noon,
> 'Right up above the mast did stand,
> No bigger than the Moon.
> Day after day, day after day,
> We stuck, no breath no motion;
> As idle as a painted ship
> Upon a painted ocean.

double-ender

a push-me-pull-you boat with identical (usually pointed) BOW and STERN.

downhaul

line used to pull a sail down a mast (if it will not fall from its own weight); see OUTHAUL.

down helm

to push the tiller to leeward (or turn the wheel the opposite way) in order to bring the BOW up into the wind.

draft (US)

see DRAUGHT.

drag

what an ANCHOR does when it is not holding.

draught (UK)

the distance between the WATERLINE and the very bottom of a vessel (the bottom of the keel in the case of a yacht). May also be expressed as the minimum depth of water in which a vessel will float. If this is, for example, 2 meters, she is said to "draw 2 meters."

draw

a sail is drawing (working) when the wind fills it.

dress

to set the sails on a yacht; more commonly used to describe a boat flying signal flags, pennants, ensigns, bunting, and anything else that flaps, as being "dressed overall."

drift

the movement of a boat caused by CURRENTS and TIDES rather than by the wind.

drogue

a device used to reduce speed in heavy weather conditions; it looks like a long canvas bucket on the end of a long rope, or even like a parachute. When hurtling downwind in heavy seas and your nerve is about to go, streaming a drogue off the STERN will cause it to act as a brake. If your nerve has gone completely (i.e., common sense has prevailed) and you are HOVE TO, a drogue can be deployed from the bow. In this case it acts as a SEA ANCHOR and keeps the BOW head-to-wind.

dynamic positioning

a computer- and GPS-based system of controlling the exact position and heading of a vessel using all available propellers and lateral thrusters; accurate enough to be able to hold a ship over an exact position on the sea bed.

ease

to ease or ease off, to slacken a line.

earings

the top corners of a square sail; not to be confused with EARRING.

earring

ornamental device in the form or a ring that goes through pierced hole in the lobe of the ear; made famous by pirates. Now made infamous by the daughter's new boyfriend. Sadly, though, he is a brilliant foredeck hand. . . . Who pointed out that, by the time you can afford to buy a comfortable seagoing yacht, you are too old to handle it without your daughter's boyfriends? Legend has it that seamen in the days of sail were allowed to wear an earring through one ear to show they had passed the Cape of Good Hope and in the other after passing Cape Horn.

east

the point of the compass at which the sun rises; in modern navigation terms, 090°, give or take VARIATION and DEVIATION.

Orient also means "east," toward where the sun rises. *Jih pun* is old Chinese for "sunrise" and the origin of "Japan." The modern Putonghua (Standard Chinese, Mandarin) for Japan is *Ri-ben*; *ri* meaning "sun" and *ben* meaning "root" or "origin." But why? If you stand on the Bund in SHANGHAI and look toward where the sun rises, you are looking toward Japan, the Land of the Rising Sun.

ebb

the withdrawal of a tide as it passes from high to low water, with the current going out in the direction of the sea. See FLOW.

echo-sounder

see DEPTH FINDER

eddy

a movement of water contrary to the direction of the main tidal stream; worth knowing about if you are RACING.

elephantiasis

see MOSQUITO.

elevation

the elevation of a CELESTIAL OBJECT is the vertical angle between that object and the horizon, subject to various corrections; used in CELESTIAL NAVIGATION.

El Niño

a warm current originating in the middle of the equatorial zone of the Pacific Ocean and then flowing south along the coast of Ecuador. Every three to eight years el Niño extends down the coast of Peru, where it can have a devastating effect on the weather. Rainfall becomes very heavy and the easterly trade winds can disappear or even veer westerly. There is a growing consensus among scientists that el Niño has become more severe as a consequence of global warming. The name is a shortening of the Spanish *El Niño de Navidad*, "the boy-child of Christmas," so-called because that is the time of year the current originates.

engines

see AUXILIARY ENGINES.

ensign

a flag flown on a vessel to signify its nationality.

Emergency Position Indicating Radio Beacon (EPIRB)

Emergency Position Indicating Radio Beacon; does what it says on the can. When this floating electronic device is activated (manually or by contact with water) it transmits a radio signal with user registration data (if you remembered to complete and mail the registration form that came with it) as well as position information to a network of satellites that assist the Coast Guard in conducting an emergency rescue. The current models all work on 406MHz frequency and operate on a worldwide basis. It is not recommended that earlier models be purchased.

Etesians

see MELTEMI.

exploring small islands and remote coastlines

Sailing is only a means to an end, a challenging and enjoyable way of getting there. Taking time off to explore remote islands and coastlines is a different kind of enjoyment. If you do decide to go ashore in some uninhabited location, plan it with care; do not put yourself or your boat at unnecessary risk. (And only make such landings once you have officially cleared immigration for the country whose territory you are on.)

Before you go off anywhere, deploy all your ANCHORS. If practical, run a couple of warps ashore. Demobilize the boat and make it as secure from intruders as you can. With a larger crew leave someone onboard.

Don't get lost! If you haven't got a map, follow this strategy. Use tracing paper to make a copy of the coastline from a chart and include conspicuous landmarks. Head in a distinct direction away from where the boat is anchored (not directly inland). This will mean that the coastline is always to your left or right. If you get lost, head for the beach and backtrack to the boat, keeping the sea in sight. If you are on an island, I can give you a 100 percent guarantee that this will work. If the island is Australia, you might wake up one morning in a doorway in King's Cross, Sidney. Take a compass with you (you should do this anyway) and add value to your sketch-map as you go along, specifically landmarks and water sources not shown on the chart. If you have a handheld GPS receiver you might take that, marking the boat's position as a WAYPOINT.

Finally, do bear in mind that if you have been at sea for some time, your legs might not be up to climbing hills. Don't be over-ambitious.

eye

1. in the context of wind, "the eye of the wind" denotes the exact direction from which the wind is originating.
2. in the context of a rotating weather system (tornado, typhoon, or hurricane) denotes the exact center, the calm center of the storm; a place you do not want to be.
3. a small loop at the end of a rope or wire; usually fashioned around a tear-drop shaped piece of metal or plastic (known as a THIMBLE), the end being spliced back on itself.

eyes

after a few weeks on your own, you can certainly begin to see things:

> Thine eyes shall behold strange women, and thine heart shall utter perverse things. Yea, thou shalt be as he that lieth down in the midst of the sea, or as he that lieth upon the top of a mast.
>
> Proverbs 23: 33–34

I think I can relate to that. Regarding the bit about lying "upon the top of a mast," though . . .

fag end

the unkempt, frayed end of a rope; the unsmoked butt of a cigarette (UK).

fairlead

a metal or wooden DECK fitting employed to guide a line (SHEET, DOWNHAUL, MOORING LINE, or whatever) in a particular direction.

fairway

a channel, usually marked by LATERAL BUOYS, used for safe inshore navigation.

fall

the end of a HALYARD that is led to a CLEAT or WINCH.

fall off

to ease off when BEATING to windward by steering more to LEEWARD (away from the wind).

fantail (US)

the overhanging STERN of a vessel; similar, but not quite the same as COUNTER.

fast

to make fast, to tie up to, to secure; particularly alongside a wharf or jetty.

fathom

obsolete measure of depth; 6ft or 1.83m. Source of expressions like "I can't fathom it" ("I can't understand it"). Originates in the Old English word for "embrace," *faedm*, on the assumption that a man's outstretched arms extended to 6ft. But you do not need to know that; modern nautical charts now have all depths marked in meters.

faungidge

a species of edible tuber found on the Indian Ocean island of Madagascar.

In 2002 the faungidge played a fascinating role in determining the truth about an 18th-century tale of shipwreck and slavery entitled *Madagascar: or Robert Drury's Journal During 15 Years' Captivity on that Island*. The hero of the piece is a young midshipman serving on an East India Company ship called the *Degrave*. In 1703 while en route to India the vessel with its 180 passengers and crew is wrecked on the southern coast of Madagascar. Most survive, the story goes, but they fall into the hands of the local Tandroy people who conscript them into their army and order them into battle with other tribes. Unhappy with their prospects, the Europeans decide to escape, upping the ante by kidnapping the Tandroy chief as they leave. The outraged locals soon recapture them and massacre the whole group, sparing only four boys including Robert Drury. After surviving eight years as a slave, Drury is finally rescued by a visiting English ship.

The book—published in 1728 and probably written by Daniel Defoe—was promoted as being a factual account of these dramatic events (a bit like the *Blair Witch Project*) but readers scoffed at this and *Madagascar* was dismissed as an elaborate fabrication. However, a few years ago, an American academic uncovered evidence that Robert Drury was a real person. Mike Parker Pearson, an archaeologist at Sheffield University in the UK, carried out further investigations in Madagascar, the highlights of which were the discovery of the wreck of the *Degrave* and being taken hostage himself by the Tandroy. The contents of Drury's book proved to be remarkably accurate; the geography of the island, the names of the tribes and their villages, and the fact that their staple food was the faungidge tuber.

feather, to

to adjust the angle of a BLADE in order to reduce its resistance to movement through the water. This can apply to an OAR when rowing or to the blades of a variable-pitch PROPELLER.

fender

a means of protecting a vessel's hull from chafing against an adjacent vessel or quay; traditionally fashioned from fancy interlaced lengths of rope, nets, or bags full or cork, old automobile tires, or (from your friendly neighborhood chandlers') smart inflated sausage-shaped tubes made from rubber or heavy-duty PVC. To go with the last are fender socks in matching colors—a serious fashion item for any boat. When you take the task in hand and use boathooks or boots to keep the other vessel clear of your paintwork, you have just "fended off."

ferrocement (ferroconcrete)

a method of hull construction using chicken wire and cement; not quite as bizarre as it seems. The basic shape of the hull is formed from stainless steel rods interwoven with multiple layers of wire mesh. The result is then coated and made watertight with a rendering of a sand-and-cement mortar. "Sounds like cruising in a culvert," was the reaction of one friend.

fetch

1. the distance to windward unobstructed by land.
2. to fetch up, to arrive or to clear somewhere or something without having had to TACK. But if you had to tack twelve days earlier for a short while, does that count?

fiberglass

see GRP

fiddle

a wooden strip or lip around any table, the purpose of which is to stop items sliding off when the boat is heeling over.

Works quite well on the navigation table; not so well on the dining table. Both hands are busy as you pour a glass of wine. Then someone places a very full

plate of CHICKEN SUPREME on rice in front of you. A gust of sou'westerly nudges the yacht, keeling her over. The plate starts sliding across the table heading for your lap. But our sturdy, craftsman-made fiddle comes into action and brings the plate up dead! Sadly though, the chicken supreme keeps going and dives over the side of the table. . . . Inevitably, I am not the first to encounter this problem. This is what Miss Emily Brittle wrote to her parents after they sent her on "a trip to the Indies . . . in pursuit of a swain."

> It was often the case on a rough squally day,
> At dinner our ship on her beam-ends would lay;
> Then tables and chairs on the floor all would jumble,
> Knives, dishes and bottles upon us would tumble;
>
> At late, when a roll brought us all to the floor,
> Whilst the ladies were screaming, the gentlemen swore,

Our Purser, as big as a bullock at least,
Lay on poor little me, like an overfed beast.

Not many weeks since, I had only to scoop
From my lap the contents of a tureen of soup;
And when with clean clothes I again had sat down,
A vile leg of mutton fell right on my gown.

Sometimes I was soiled from my head to my toe
With nasty pork chops, or a greasy pilau.
Full many a glass of good wine, I may say,
By a violent toss was thrown down the wrong way;

And, as on board ship we have no one to scrub,
As for three months at least there's no thumping the tub,
So I think it but proper that *delicate* women
Should lay in a plentiful stock of clean linen.

from *The India Guide; or, Journal of a Voyage to the East Indies in the Year MDCCLXXX*
by Sir George Dallas (1758–1833)

This problem also applies to those non-slip placemats that you can buy in chandleries. Maybe someone should try making plates from the same stuff? To keep the food stuck on? Or use bowls? Fiddles are also fitted across the front of bookshelves and this arrangement works fine; I have never had chicken supreme fall into my lap from a bookcase.

fiddle block

a BLOCK with two SHEAVES in alignment, but with one larger in diameter than the other.

film canisters

the type of canisters used by Kodak to hold 35mm film cassettes are worth saving as a means of keeping small objects and substances dry onboard. For example, cooking spices usually come in small jars (which can break) or in cardboard cartons (which self-destruct, destroying the contents if they get wet). Tip: Transfer each herb or spice to film canisters and label the top and side using white typewriter correction fluid (such as Wite-Out®) to avoid confusion (especially on the part of customs and the DEA).

fin keel

see KEEL.

fire

fire onboard a boat is probably a bigger danger than sinking. Fires can of course be tackled with fire extinguishers but the real nightmare is the risk of naked flames reaching explosive materials such as propane gas or petrol (gasoline). These should be stored in DECK lockers that drain overboard.

firearms

Some, not many, cruising yachtsmen carry guns onboard, primarily for personal protection. Eric Hiscock, the doyen of ocean passage making, kept a shotgun onboard *Wanderer IV*. Captain Joshua SLOCUM was also "tooled-up" (as they say in the London criminal underworld) when he arrived at Punta Arenas for his passage through the backwaters of Tierra del Fuego. The locals were both fascinated and appalled at his plan to make the trip single-handed.

> The port captain, a Chilean naval officer, advised me to ship hands to fight Indians in the strait farther west, and spoke of my stopping until a gunboat should be going through, which would give me a tow. After canvassing the place, however, I found only one man willing to embark, and he on condition that I should ship another "mon and a doog." But as no one else was willing to come along, and as I drew the line at dogs, I said no more about the matter, but simply loaded my guns. At this point in my dilemma Captain Pedro Samblich, a good Austrian of large experience, coming along, gave me a bag of carpet-tacks, worth more than all the fighting men and dogs Tierra del Fuego. I protested that I had no use of carpet tacks on board. Samblich smiled at my want of experience and maintained stoutly that I would have use for them. "You must use them with discretion," he said; "that is to say, don't step on them yourself." With this remote hint about the use of the tacks I got on all right, and saw the way to maintain clear decks at night without the care of watching.

Later, sailing out of an ANCHORAGE at Cove Island, SLOCUM had his first encounter with the "Indians."

> Canoes manned by savages from Fortescue now came in pursuit. The wind falling light, they gained on me rapidly till coming within hail, when they ceased paddling, and a bow-legged savage stood up and called to me, "Yammerschooner! yammerschooner!" which is their begging term. I said, "No!" Now, I was not for letting on that I was alone, so I stepped into the cabin, and, passing through the hold, came out at the fore-scuttle, changing my clothes as I went along. That made two men. Then the pieces of bowsprit which I

> had sawed off at Buenos Aires, and which I still had on board, I arranged forward on the lookout, dressed as a seaman, attaching a line by which I could pull it into motion. That made three of us, and we didn't want to "yammerschooner!; but for all that the savages came on faster than before. I saw that besides four at the paddles in the canoe nearest to me, there were others in the bottom, and that they were shifting hands often. At eighty yards I fired a shot across the bows of the nearest canoe, at which they all stopped, but only for a moment. Seeing that they persisted in coming nearer, I fired the second shot so close to the chap who wanted to "yammerschooner" that he changed his mind quickly enough and bellowed with fear, "Bueno jo via Isla," and sitting down in his canoe, he rubbed his starboard cat-head for some time. I was thinking of the good port captain's advice when I pulled the trigger, and must have aimed pretty straight; however, a miss was as good as a mile for Mr. "Black Pedro," as he it was, and no other, a leader in several bloody massacres. He made for the island now, and the others followed him. I knew by his Spanish lingo and by his full beard that he was the villain I have named, a renegade mongrel, and the worst murderer in Tierra del Fuego. The authorities have been in search of him for two years. The Fuegians are not bearded.

Today, you are much more likely to be boarded by a team of fresh-faced young chaps from a US Coast Guard cutter. As they will be (a) nervous and (b) armed with M4 assault rifles, you would be ill advised to brandish anything that looks remotely like a firearm. And even in the waters of Southeast Asia, producing a shotgun would be a fast way of learning about the superior firepower of the AK-47 Kalashnikov.

Probably the main reason for not carrying guns on boats is that—even if they are licensed—they will be removed by port authorities on arrival and kept locked up until you leave. The result is a lot more paperwork and a lot less security than you imagined. But there are other reasons. In an Internet newsgroup discussion of the matter, someone pointed out, "An accidental discharge on board could put a hole in your bottom." To which came the speedy response, "But I've already got a hole in my bottom!"

first aid and medication

When at sea, even when only a few miles offshore, you are no longer within easy reach of the emergency services. This makes it essential for serious seafarers to learn first aid—a crew member's life might depend on it. There are many approved courses available including some specific to first aid at sea. In addition to the training, there needs to be at least one good first aid kit onboard and, in the case of a passage making yacht, a selection of essential pharmaceuticals. I also recommend a smaller first aid kit be placed close to hand in the cockpit. Here are some recommended contents.

Cockpit First Aid Kit			
Dressings	Large wound dressings	1	
	Medium wound dressings	1	
	Crepe bandage 7.5cm	1	
	Assorted adhesive bandages	1	Small box
	Antiseptic wipes (in sachets)	2	
	Sterile gauze swabs	1	Packets
Instruments	Dressing scissors	1	Pair
Medication	Antiseptic cream	1	Tube
	Betadine solution 100ml	1	Plastic bottle
	Aspirin/acetaminophen	25	Tablets

PACKAGING AND STORAGE

- Must be kept in a strong watertight container.
- Store in a cockpit locker or on the bulkhead just inside the cabin so you can reach it from the cockpit (next to the VHF mike). If this is also close by the galley, so much the better.

This could also serve as the first aid kit you take ashore. See EXPLORING SMALL ISLANDS AND REMOTE COASTLINES.

If buying a proprietary First Aid Kit, use this checklist to ensure that the contents are comprehensive and likely to meet most if not all of your requirements. The quantities of each item will obviously depend on the number of crew and length of voyage. Ensure that the kit is packaged in a good watertight container and includes an instruction booklet. *Keep all sharps sealed until needed! Customs officers can be very suspicious of nomadic yachties carrying syringes!* This is most likely to apply if you have an injecting diabetic onboard.

Main First Aid Kit			
Dressings	Triangular bandage	3	For slings, etc.
	Cotton wool 50g		
	Large wound dressings	2	
	Medium wound dressings	3	
	Conforming bandages	3	
	Crepe bandage 7.5cm	1	
	Adhesive waterproof strapping	1	2.5cm roll
	Assorted adhesive bandages	1	Large box
	Eye pad	1	
	Antiseptic wipes (in sachets)	10	
	Sterile suture strips	1	Packet
	Sterile gauze swabs	2	Packets
	Kerosene gauze dressings	10	
	Safety pins	10	
	Fingerstalls, size 3 and 8	2	
	Isopropyl alcohol 70%	1	100ml plastic bottle
Instruments	Dressing scissors	1	Pair
	Tweezers	1	Pair
	Medical thermometer	1	
	Disposable scalpel	1	With replacement blades
	Safar tube	1	
	Disposable catheter	1	
	Disposable syringes 5ml	10	
	Disposable syringes	5	

PHARMACEUTICAL KIT

If you are planning to spend a lot of time at sea, an onboard pharmacy is a good idea. Consult with your doctor about this list; you need to explain what you are doing because he will have to write prescriptions for some of the medication. Tell him or her about the PAN-PAN MEDICO system and that in the case of serious injuries or ailments there will be a doctor on the other end of the radio telling you what to do.

Pharmaceutical Kit		*Prescription needed?*
Amoxycillin 500mg capsules*	Infections of ear, nose, throat, mouth, air passages	Yes
Aqueous Cream (emollient cream)*	Dry and chapped skin	No
Asilone antacid tablets	Gastric complaints, stomach ulcers	No
Benylin	Chesty coughs	No
Bisacodyl	Constipation (see also Senna)	No
Brolene Ointment	Severe conjunctivitis (caused by infection), styes	No
Buscopan tablets, ampoules	Colic, cramps (gallbladder, kidneys)	Yes
Claritin	Allergies, itching (this antihistamine does not cause drowsiness)	No
Co-codamol	Suitable for moderate pain	No
Cyklokapron	Internal bleeding of stomach, intestines, bladder	Yes
DF118 elixir and ampoules (Dihydrocodeine)*	Severe pain. Patient should avoid alcohol during this treatment!	Yes
Diclofenac 50mg*	Inflammation, swelling. Is also an effective pain-killer in the case of injuries. NOT to be taken by anyone with stomach ulcers.	Yes
Germoline cream	Boils	No
Glyceryl trinitrate tablets*	Angina. However, an angina patient should be carrying their own medicine.	No
Hydrocortisone cream*	Severe sunburn, insect bites, eczema and dermatitis	No
Lasonil cream	Sprains, bruises, contusions, piles	No
Loperamide*	Severe diarrhea and stomach cramps	No
Meclozine tablets	Seasickness	No
Motilium	Severe vomiting and nausea	No
Opticrom (Sodium cromoglycate)*	Conjunctivitis caused by allergy	No
Otrivine Spray (EU excluding UK)	Colds	No
Paracetamol	Headaches and other pain	No
Penbritin injections 250mg (ampoules and capsules)	Infections of urinary tract, bile duct (capsules), blood poisoning (ampoules)	Yes
Rhinospray (not available in UK)	Colds	No
Sandocal (calcium) tablets	Allergies, itching	No

Pharmaceutical Kit		*Prescription needed?*
Savlon cream	Burns, sunburn, grazes, insect bites	No
Savlon liquid 500ml	Disinfection of wounds	No
Senna	Constipation. Senna is much more widely available than Bisacodyl.	No
Stemetil suppositories 5mg	Severe vomiting	Yes
Tyrozets	Sore throat	No
Valium Tablets 2mg	Nervousness, insomnia	Yes
Zantac Tablets	Chronic or severe dyspepsia	No
***Generic name**	**WARNING: Medicines rarely mix well with alcohol. Step 1 with any patient should always be a ban on the booze until they are fit.**	

After consulting our doctor, we also consulted our lawyer, who insisted we point out that neither the author nor the publisher can accept responsibility for the consequences of any of the following drugs being misused.

Check "use-by" date and condition of medicines regularly. Signs of deterioration include discoloration, loss of form (tablets disintegrating, precipitation or coagulation of liquids), and change in smell. Some preparations (e.g., penicillins, insulin) keep better when refrigerated.

There are two books that are worth having onboard; a general health guide and a reference book to pharmaceuticals. I recommend *The BMA Complete Family Health Guide* by Dr. Tony Smith (Dorling Kindersley); the same publisher also has a CD-Rom called *Family Health Encyclopedia*. And the *BMA New Guide to Medicines and Drugs* (Dorling Kindersley again) has all you need on more than 2,500 prescription and over-the-counter remedies. There is a more compact version called *The Concise Guide to Medicines and Drugs*.

STERILE MEDICAL KIT

The following items are for use by professional medical staff only. They are obtainable in a sealed pack and the list below is based on the contents of the Dixon Community Care's "Sterikit" (UK). They are recommended for yachts visiting countries where the sterile condition of instruments cannot be guaranteed (for example, the reuse of hypodermic needles).

Sterile Medical Kit			
Mini Vein	1	Syringes 5ml	3
Blood Giving Set	1	Needles 21g	5
Medi Swabs	4	Silk Suture 50cm x 26cm	1
Strip Skin Closures	8	Stitch Cutter	1
Melolin Dressings	2	Antiseptic Tissues	6
IV Cannula Catheta	1	Microporous Tape 1.25cm	1

Finally, make sure you include in your first aid kit blister bandages of the type sold under the brand name of Compeed. Now I have never known anyone to get blisters on a boat, but these Compeed things are nothing like regular Band-Aids. The adhesive is very durable, they are thin, flexible, and waterproof—just what you need for an emergency repair to any material short of marine stainless steel. They are an excellent quick fix for a snag in waterproof clothing, broken spectacles and, of course, blisters. Not quite as vicious as DUCT TAPE.

fish, to

1. to fish a spar means to repair it by attaching one or more pieces of wood (known as a fish), usually by binding on with line or cord; the result resembles a splint on a broken leg.
2. to lift an ANCHOR completely onto the DECK.
3. to supplement one's diet by catching the fruit of the sea; excludes the acquisition of flying fish that have a charming propensity to surrender by crash-landing onto the DECK; this is usually because they are being chased by voracious TUNA.

> So is this great and wide sea, wherein are things creeping innumerable, both small and great beasts.
> There go the ships: there is that leviathan, whom thou hast made to play therein.
> These wait all upon thee; that thou mayest give them their meat in due season.
> That thou givest them they gather: thou openest thine hand, they are filled with good.
>
> Psalm 104, King James Bible

See TROLLING, KILLING FISH, FISHING AT ANCHOR.

Fisherman's Bend

a knot similar to the ROUND TURN WITH TWO HALF HITCHES; in this case, the half hitch is passed under the bend so that it is gripped under load; effective for securing a line to a rail or ring; also known as the ANCHOR BEND.

fishhook (US)

slang for any piece of rigging that injures you unexpectedly (e.g., a frayed metal cable on a lifeline or shroud).

fishing at anchor

If you are becalmed or anchored up in an exotic lagoon somewhere, you might see some decent-sized fish swimming close to the yacht. Try using a gaff to catch these. (That's "gaff" as in a pole with a sharp hook at the end, and not "gaff" as in sailboat with a four-cornered MAINSAIL, of course.) Attach a line to the gaff so you do not lose it in the excitement. Attach a line to yourself, too—I would feel really bad if anyone went overboard merely because they were trying out one of my mad fishing ideas. I would feel even worse if you were unable to get back onboard. An alternative in still waters is to use a hunting bow. This is the type of bow used by hunters to kill Bambi's mother, but using barbed arrowheads with a light line running back to the boat.

fitting-out

the ritual of making a vessel ready for sea; to overhaul.

fix

a fix is the vessel's position determined by means of BEARINGS of MARKS and/or RADIO BEACONS, CELESTIAL OBJECTS, or GPS; note that a position obtained by DEAD RECKONING is *not* a fix but a fix will confirm a dead reckoning position.

fixed stars

stars that move very little (over a very long period of time) in relation to each other. Without fixed stars there would be no CONSTELLATIONS and without constellations it would be impossible to identify the individual stars we need for CELESTIAL NAVIGATION.

flags

see SIGNAL FLAGS.

Flags of Convenience

a heinous legal device invented by the shipping industry to increase its profits by registering ships in countries with less than stringent safety regulations; has the effect of dumping experienced and well-trained seamen and officers in favor of anyone who is prepared to go to sea for lower pay. Results in putting at risk the lives of everyone who goes to sea including the rescue services and, in the case of bulk carriers, can cause significant risk to the environment. How the INTERNATIONAL MARITIME ORGANIZATION can allow such an aberration is beyond me—it certainly wouldn't be allowed in the aviation business.

flake

1. to fold a sail in tidy layers on a boom; but see LAZYJACKS.
2. to coil a rope or chain in such a way (e.g., in figures-of-eight) that it can be run out without fouling; the effective flaking of an anchor WARP on the FOREDECK will enable an ANCHOR to drop without taking a tangled lump of chain (or your leg) with it.

flap

flags "flap" in the wind (sails FLOG). When flag signals were being sent back and forth during the heat of battle, there was said to be "a flap on"; in other words a sense of urgency just short of panic.

flare

1. the upward and outward curve of the bows and/or upper sides of a vessel; the opposite of TUMBLE-HOME.
2. "flares" are pyrotechnic DISTRESS SIGNALS; may be handheld, fired into the air as a rocket, or shot from a VERY PISTOL. There are various approved types as shown below.

Orange smoke	Handheld or floating	Smoke is used to attract attention, but is ineffective at night. Extremely useful to deploy when being rescued by helicopter as the smoke provides the pilot with an indication of wind direction, strength, and variability.
White flares	Handheld	Used to indicate your position.
Red flares	Handheld or fired by rocket or flare pistol	Used in Mayday situations. There is a more expensive version which descends by parachute thereby extending its over-the-horizon visibility.

Great care should be taken when using flares. Use up recently outdated stock to practice but get clearance from the COAST GUARD first on Channel 16 before firing them off.

I once had to dispose of some *very* old stock (more than ten years out of date) so I stuck them in a shopping bag and headed through the midday heat for nearest US Coast Guard Station. It turned out to be an auxiliary squadron and the place was closed. But not far away was an Army National Guard post. I walked in the front door but there was no one around. I headed left down a corridor leading off the entrance hall and peeked in through the open doors of offices stacked with top-secret documents but the place was as deserted as the *Mary Celeste*. I reversed tracks, my bag of explosives swinging by my side, and followed a reciprocal course past even more empty-but-open offices. Eventually I heard voices and walked in on a crowd of guys in uniforms, boots on desks, eating donuts; it was lunchtime. I showed them the flares and explained that boaters are supposed to give their expired pyrotechnics to the Coast Guard rather than put them in the garbage or save them for July 4th. The sergeant who was smoking emptied the bag onto his desk and juggled with a couple of the rocket flares. I thanked them, made my excuses, and beat a hasty retreat.

flare gun

a pistol for firing flares into the sky, sometimes called a VERY PISTOL; see FLARE (2).

flashlight (UK: torch)

A handheld source of light. In my experience it is essential to have at least one of these onboard. These should be (a) a conventional handheld job the best of which by far is the Maglite®, the only make I have seen that looks properly designed and engineered and well worth the extra cost, (b) a backup for (a), (c) the type that fits to a headband for use when you need your hands free (e.g., to change an IMPELLER), and (d) a powerful lamp for use as a headlight in poor light and as a signaling lamp. Powered from the 12v supply (and sometimes rechargeable) these are now rated at up to 1,000,000 candlepower. Make sure you get one with a Morse key for signaling to other vessels. Make sure you know Morse.

Tip: If you wish to draw attention to yourself, shine the light onto your (white) MAINSAIL.

flattie, a

a dinghy or larger DAYBOAT with a flat bottom and CHINES.

flog

when a sail flaps uncontrollably it is said to "flog."

flood

a flood tide is one that is going from low to high water; an incoming tidal current.

flotsam

anything found floating on the sea, some of which may be JETSAM.

flow

the onset or ingress of a TIDE as it passes from low to high water, with the current going in from the direction of the sea; a flood tide. See also EBB, NEAP TIDE, and SPRING TIDE.

fluke

the flat pointed plates attached to the head of an ANCHOR and that bite into or hook onto the sea bottom causing the anchor to take hold and secure the vessel; see ANCHORS.

fluky

used to describe a light, variable wind.

flush deck

a deck that runs continuously from fore to aft.

fly

the size of a flag measured along its horizontal edge.

flying bridge

elevated upper-story HELM station, located above the main helm at DECK level; most commonly seen on sport fishing boats.

flying jib

an extra sail set beyond the JIB rigged to a BOWSPRIT extension and the head of the MAINMAST. A flying jib is hanked to the forestay.

FOA

Flies Open on Arrival; a major cause of death recorded by the RNLI after performing the unpleasant task of recovering a body from the sea. I blame the HEADS. This dire piece of plumbing is enough to drive anyone to standing in the lee pushpit and peeing over the side.

fog

there are two types of fog; ADVECTION FOG and CONVECTION FOG.

fog signals

a system of audible signals using horns ("whistles"), bells, and gongs when visibility is restricted; described in the COLREGS (US: NAVRULES). On a small yacht (less than 12m), you just need a horn.

Folkboat

a 25-foot LOA sailing yacht of cruiser-racing design built of mahogany strip plank construction and originating from Scandinavia.

following sea

when the waves are coming at you from the STERN, you are in a following sea.

foot

the bottom edge of any sail.

footloose

see LOOSE-FOOTED.

fore

to the front; opposite of AFT.

fore-and-aft

in the direction of a line drawn between the BOW and STERN of a vessel.

forecabin

accommodation area in the BOW of a yacht; introduced by designers to assuage the owners unhappy with the demise of the FORECASTLE; the CHAIN LOCKER, lots of wet sailbags, and (on smaller yachts) a HEAD (HEADS) also share this space. It is perfect, though, for two crew who like to play footsie in bed.

forecastle (pron. "fo'c'sle")

a shack built onto the FOREDECK of traditional sailing ships; normally used to house the crew in splendid isolation. (This is an arrangement some skippers would love to see reintroduced.) It is the place where Ishmael used to lay his head:

> No, when I go to sea, I go as a simple sailor, right before the mast, plumb down into the forecastle, aloft there to the royal mast-head. True, they rather order me about some, and make me jump from spar to spar, like a grasshopper in a May meadow. And at first, this sort of thing is unpleasant enough. It touches one's sense of honour, particularly if you come of an old established family in the land, the Van Rensselaers, or Randolphs, or Hardicanutes. And more than all, if just previous to putting your hand into the tar-pot, you have been lording it as a country schoolmaster, making the tallest boys stand in awe of you. The transition is a keen one, I assure you, from a schoolmaster to a sailor, and requires a strong decoction of Seneca and the Stoics to enable you to grin and bear it. But even this wears off in time.
>
> *Moby Dick; Or The Whale* by Herman Melville, 1851

foredeck

area of a vessel at the BOW; in the case of a yacht, between the MAINMAST and the PULPIT.

forefoot, the

the BOW between the CUTWATER and the leading edge of the KEEL.

forepeak

the below-deck space at the BOW; often used in a sailing boat to contain a manual or powered CAPSTAN.

foresail (pron. "fors'l")

any sail set forward of the MAINMAST.

fore-staff

See CROSS-STAFF.

forestay

a STAY that runs from the STEM of a yacht to the top of the foremast (or, in the case of a FRACTIONAL RIG, to a point about two-thirds of the height of the mast); the JIB is usually HANKED to this stay.

forward (pron. "forrard")

anywhere on a vessel nearer the BOW than the STERN or where you are standing at the time.

foul

used to describe a rocky seabed that will cause an anchor to snag and set incorrectly.

fouled

snagged, caught up, or twisted; usually said of a line or an ANCHOR.

foulies (US)

foul weather gear (waterproof clothes); see also OILSKINS.

founder, to

to sink. The word stems from the Latin *fundus* for deep (also found in "profound"). The Old French verb *fondrer* means "to submerge."

frapping turns

the turns of a rope that serve to secure objects together.

free, sailing free

when a yacht is not CLOSE-HAULED, she is said to be "sailing free."

freeboard

the distance between the WATERLINE and the DECK; always seems to be a terrifying height when viewed from the water.

frigate

[from the French *frégate*]

a general-purpose warship carrying various armament; in the US Navy a frigate is bigger than a DESTROYER, in the Royal Navy a frigate is smaller than a destroyer.

In the 18th century frigates were three-masted, squared-rigged ships that compromised the heavy fire-power of the "ships of the line" in favor of speed and agility. Nelson's flagship HMS *Victory* sported 104 heavy cannon on three decks; thirty of these were 32-pounders (that's the weight of the cannonballs, not the guns of course). The *Dolphin* made do with 24 guns on a single deck and these were pretty lightweight 9-pounders. Her length overall was 34 meters and 10 meters in the beam; she was crewed by 160 sailors.

full and by

sailing as close to the wind as possible while continuing to keep the sails at maximum pulling power (no wrinkling at the LUFF of the MAINSAIL); also sailing by the wind rather than to a COURSE.

full and change

the moon is "full and change," a new moon.

furl, to

to roll or gather up a sail and secure them to the boom (using GASKETS).

futtocks

a structural timber frame made up from several pieces of wood (laid end to end) and numbered 1st futtock, 2nd futtock, etc., onto which the outer planking is fixed. They are situated in the middle part of the ship between the floor and top timbers. What a wonderful word!

G

gadget

A handy word for a "gubbins" or a "doojigger." It is included here because it seems to have originated as seaman's slang. It is a relatively new word, first seen in print in the latter half of the 19th century:

> Then the names of all the other things on board a ship! I don't know half of them yet; even the sailors forget at times, and if the exact name of anything they want happens to slip from their memory, they call it a chicken-fixing, or a gadjet, or a gill-guy, or a timmey-noggy, or a wim-wom—just pro tem, you know.
>
> from *Spun Yarn and Spindrift* by R. Brown, 1886
> (Quoted in *Dictionary of Word Origins* by John Ayto)

Better than calling a "thingummybob" a "howsyourfather," I suppose.

gaff

1. a SPAR attached to the head of a four-sided gaff sail in a FORE-AND-AFT rig and hauled up the mast by HALYARD when the sail is set.
2. a pole that ends in a sharp hook; used by fishermen to haul large fish over the side.

gaff jaws

the U-shaped fitting that engages the mast at the inboard end of a GAFF; enables the sail to move in relation to the CENTER-LINE and the gaff to move up and down the mast.

gaff sail

four-sided sail used as the MAINSAIL in a GAFF rig

Galapagos Islands (and Three-in-a-Bed Sex Romps)

This is the second of three entries that tell the story of a 1930s passage from Denmark to the Pacific Ocean. Part 1 is under MONSOON and Part 3 under VANIKORO SHIPWRECK.

The Galapagos is the *Monsoon*'s next stop after leaving the Panama Canal en route to the South Pacific. You may have thought it was the animal wildlife that was a little curious about the Galapagos. But that was before the German aristocracy turned up. In the 1930s there were strange goings-on within the tiny community of Europeans on the islands—especially on Floreana. Tabloid headlines would have splashed "Three-in-a-Bed Sex Romps on Desert Isle!" and "Sex-mad Baroness in Death Plot!" (All right, you and I know the Galapagos are not desert islands, but tabloid newspapers have never let the facts get in the way of a great story.)

According to the account in my possession, there once stood a sign on the beach by the main landing-spot. This is what was written on it:

WHOEVER YOU ARE—FRIENDS!

Two hours from here is the hacienda "Paradise." It is a spot where the tired traveller has the happiness to find peace, refreshment and quiet on his way through life.

Life—this small portion of eternity which is bound to a clock, is so short—so let us then be happy—let us be good!

In Paradise you only have one name—Friend!

With you we will share the salt of the sea, the vegetables of our garden and the fruit of our trees, the cold water which runs down our cliffs, and the good things friends brought us when they passed this way.

We will spend one or two moments of life with you and give you the happiness and peace that God planted in our hearts and souls when we left the restless metropolis and journeyed away to the quiet of the ages, which has spread its cloak over the Galapagos.

(Signed) BARONESS WAGNER-BOUSQUET

One day in the mid-1930s a handsome young Dane stood before the sign, reading it with bemused curiosity. It was our chum Hakon Mielche from the schooner *Monsoon* (see MONSOON, VANIKORO SHIPWRECK) and he had a letter from the Governor on Chatham Island to deliver to the aforementioned Baroness. He set off up the hill. This is his account of what happened next.

It turned out that in the two hours mentioned in the notice one was supposed to cover exactly ten miles and the marquis gave no shadow. It was on a large plateau scorched by the sun that I found the first signs of the presence of man. The skeletons of a couple of wild cows lay there. At first glance I thought that it was the work of roving wild dogs, but a closer investigation revealed that both thighs and ribs were missing from the assembly of bones, and that in the middle of the forehead there was a hole—the result of a dum-dum bullet.

The path from the plateau was marked all along its length by perforated skulls, from which one might conclude that the Baroness and her archangel lived on something other than cold water and the fruits of the tree. The road wound on across plateaux, through passes and over rocks, and seemed as if it would never come to an end. From the position of the sun the business had lasted too long for me already, but, finally, I did reach my journey's end, with my tongue hanging out and a sucked lemon between my lips—if the reader can imagine such a combination.

At the end of the path stood an unhappy and curiously out-of-place Japanese gate with "Welcome" painted on it in vulgarly large letters

> which bored into my eyes that were already smarting from the sun. A donkey hee-hawed somewhere behind the gate; and there came a spirit lightly tripping over the stones—the Baroness von and zu Wagner-Bousquet and Floreana.
>
> I fought heroically with my sense of humour and sent it flying into the bushes with a well-aimed upper-cut before I dared look at her again. She stood beside me, apparently not at all surprised at my coming. She tendered me a little white hand, and we glided into Paradise, past the local St. Peter who hee-hawed his blessing behind us.
>
> The Baroness was small, but one could not say that she was beautiful. In front of her swollen lids she wore strong spectacles and her mouth, though too large, was yet unable to cover her long, yellow, rabbit teeth. [I blame the in-breeding of Europe's aristocracy—JL.] Her hanks of hair were kept in place by means of a pink shoulder-strap around her head, and she wore a kind of baby's rompers, like the trunks the ladies of the chorus wear when rehearsing. She moved in that hopping manner which jockeys call a "canter."
>
> I cantered beside her as well as I could, till, by way of a change and to show a little personal character and independence, I tried one or two chassé steps, but unfortunately stumbled over a large lump of lava.
>
> Such was our entry into the hacienda "Paradise," a wooden hut set in the middle of a vegetable garden, where a powerfully-built, blond youth gave me a paw and was introduced by the Baroness as "My Baby."
>
> Baby looked as though he had been a gigolo in a very cheap restaurant somewhere in Berlin. His eyes were a watery blue, his hair was curly and his smile much too sweet. In private life he was Herr Rudolf Philpson of Berlin, aged 28. A German cook, tubercular and with one foot in the grave, smiled in a sickly way from the background and brought tea.

The Baroness, Hakon learned, had been born about forty years previously in Austria. She'd spent a lot of her youth in the Middle East, where her father supervised the building of the Baghdad railway. In Syria she met "a daring young French" air force pilot. Then she met "Baby"—Rudolf Philpson. "Shortly after the sky went up in flames, as the aviator was not modern enough to tolerate a lover in his household, and a divorce was put through." But the lovers soon grew tired of Paris (as you do) and they "decided to find their way back to the bosom of nature. They spun a globe on its pedestal and, stopped it with a finger and lo! a little, red varnished nail pointed exactly to the island of Floreana in the Galapagos Group." Remote the trio may have been, but not beyond the reach of the news-hounds. In the early 1930s the story of the Baroness began to appear in newspapers throughout the world.

A woman was supposed to have made herself ruler of the Galapagos, to have proclaimed herself Empress of the Pacific, waged war with Ecuador, and with her hordes of indomitable free-booters made the waters of that sea unsafe. Even the old [London] Times fell into the trap and a Copenhagen newspaper turned it into a front-page item. Some young Greek idealists ("a hundred lovers of beauty and freedom"), formed a league, and offered to help her in her struggle. There wasn't a grain of truth in it, but whoever the Baroness's publicity agent was, he knew his business.

The Americans were obsessed with her. She was the correspondent of several papers in the States, and when the millionaires' yachts brought the curious to see the lonely Queen of Floreana she was showered with gifts.

But the Baroness was not the first to "colonize" Floreana, nor was she the oddest inhabitant. After taking tea at Paradise, Hakon headed off, more mail tucked into his satchel, to see Dr. and Mrs. Ritter.

> [Dr Ritter] lived a couple of miles away from the Baroness and a notice at his gate said that one should call loudly once or twice before entering. I roared like a foghorn and there was a rustling in the bushes. Dr and Mrs Ritter were exponents of the nudist cult, but not exhibitionists. They received you in more suitable clothing and once you had seen them you were glad. . . .
>
> The Ritters were the first to come to the island. They constituted the island's ancient aristocracy and were furious with the Baroness, the parvenu, for taking away half their fame as hermits and the greater proportion of the Americans. . . .
>
> Ritter was a philosopher. He was fairly small, his legs had been screwed on wrong, so that his toes pointed inwards. His nose was long and pointed, he had watery protruding eyes and the hair of a prophet. His disciple, Miss Dora, smiled a toothless welcome. The couple had at their disposal only one pair of false teeth and this was Ritter's day.
>
> Miss Dora wore beach pyjamas and had large, naked, black feet. Her neck had not been washed for at least a month, and had been given a marbled effect by the passage of drops of sweat.
>
> Nudism is above all a healthy movement, but the Galapagos are so short of water!

> Before his exile, Dr Ritter had been a dentist in Berlin. He had married an opera singer who had had no appreciation of subtle philosophies, and then little Miss Dora came across his path. She understood him. They moved together to the Galapagos, although Brunhilde would not consent to a divorce. So now he was expecting that she would arrive one fine day on Lohengrin's swan and demand him back. Her longing for him would drive her to it—or so he said.

One can only hope Brunhilde turns up on a day when the good doctor has the teeth. But after Hakon and his friends left the Galapagos, the Floreana Madness Index rose alarmingly. The Danes only learned about the bizarre developments long after they'd returned to Copenhagen.

> The letter was from Captain Alan Hancock who has once again visited the Galapagos in the *Velero*. He had visited the colonists of "Paradise" and "Eden" [the Ritters' place] and this time in tense excitement, for Dr Ritter had sent him a letter asking him to come as quickly as possible, as things had happened and would happen, which were too terrible to be the subject of a letter.
>
> Hancock came, but the evening before his ship anchored in Post Office Bay the philosopher died of poisoning. Miss Dora sat in their primitive hut, half out of her mind with fright and grief and the story she told was like the feverish fantasies of an overstrung mind.
>
> One evening Lorenz [the Baroness's "cook"] had knocked at the door, been admitted and had begged to be allowed to stay. The night before this, the Wittmer family [another group of German colonists] had been awaked by wild howls and shrieks coming from the Baroness's property, but had not paid much attention to it as quarrels and dramatic jealousies were nothing unusual in the Baroness's household. Every now and again the "Pirate Queen" would simply order her big Baby to hunt the tubercular cook, who was already marked down for death, over the rocks and stones till he finally fell down and unresistingly allowed himself to be beaten, while the Baroness looked on and encouraged her gladiator with wild shrieks. On this evening Lorenz came to the Ritters quaking all over and begged to be allowed to stay till he could get a ship to Ecuador.
>
> According to his story, on the morning after the Wittmers had heard the wild cries, the Baroness and Baby had disappeared completely. A yacht had put in at night, they had taken their belongings and left Lorenz to his fate.
>
> Lorenz stayed quite a short time with the Ritters, then one day Nuggeröd, our pilot from Santa Cruz, ran into Post Office Bay in his little "Dynamit" and took him on board. They never reached

Santa Cruz, but for the next three weeks their boat was sighted from several other islands sailing now here, now there, seemingly without aim or purpose, until it disappeared completely.

Captain Hancock looked up the Wittmers, but they could not tell him anything he had not already learned from Miss Dora. He took her away with him to the mainland and then continued his scientific investigations among the other islands. On one small, deserted, volcanic island he found Nuggeröd and Lorenz. They lay a short distance from the shore, their bodies dried up from the sun, but still not so unrecognisable that they could not be easily identified. Of Nuggeröd's native sailor there was no trace, nor could the remains of the boat be found.

Captain Hancock had to return without clearing the matter up and the Floreana mystery is today presumably still open to conjecture. Hancock himself thinks he has built up the right solution.

In other parts of the group nothing had either been heard or seen of the mysterious yacht that was supposed to have taken off the Baroness and Baby. Nor had news been received of these two originals from other parts of the world, although a search was made for them after Hancock had got in touch with the nearest mainland. This brought the Captain to the conclusion that Lorenz had finally had enough of the thousand and one little tortures to which he was subjected, that his rage got the upper hand and he had murdered the pair in their sleep, burying the bodies and inventing the story of the yacht.

It is, however, possible that there was something more than jealousy and the desire for revenge at the back of the murder.

The Galapagos have always been a pirates' nest and the air is still thick with tales of buried treasure. Was it not possible that Lorenz overheard a conversation between the Baroness and Baby and thought that he had found out where such treasure lay buried? That he then killed his two competitors in cold blood and came to an agreement with Nuggeröd to go treasure-hunting among the islands?

That would account for the curious voyages of the Dynamit which were observed by the other settlers during those three weeks and would also explain why the three inhabitants of the boat landed on an island which otherwise had nothing of interest to offer. Did Nuggeröd and Lorenz then quarrel about the imagined treasure and kill each other, or were they wrecked and driven ashore on the island? If so, where is the boat and the native sailor? It is not impossible that he stole both boat and secret and abandoned his employer to die of thirst under the merciless sun! Then where is he now?

The whole affair is shrouded in a veil of impenetrable mystery. Ritter could perhaps have given us the key to the puzzle, but he died the day before he should have spoken and Miss Dora kept on repeating that she did not know what it was that Ritter wanted to tell Hancock, the news that was so revolting that he could not confide it to paper.

What position did Ritter take in the matter? How much did he know of the murder and the imagined treasure? and why must he die at such a suitable time for him or those who desired silence?

One question piles up on another, the threads grow confused, and the scattered, uncertain material that is at one's disposal only permits of guesses.

Certainty will surely never be had unless the Baroness and Baby turn up one fine day in some other part of the world and clear Lorenz of a suspicion which undeniably rests on him. But whatever he may have done, he has atoned for it on a block of lava under the terrible equatorial sun.

Excerpts from *Let's See If The World Is Round* by Hakon Mielche, 1944 (originally published in Danish, 1938, now out of print). The two chapters (including Hakon's drawings) describing the visit to the Galapagos have been posted online at www.galapagos.to/texts/mielche.htm.

I think Hakon missed something. Those dentures had something to do with it.

Galileo Galilei

Galileo Galilei (1564–1642) an Italian scientist and pioneer of scientific astronomy, especially the use of the telescope.

Galileo Positioning System

the name given to a European Union and European Space Agency joint venture to establish a new satellite navigation system. The target date was originally set for 2008 but, with six satellites in orbit at the time of writing (May 2013) it seems likely that the current 2018 target will be met. In addition to the twenty-seven member countries of the European Union, the Galileo positioning system is now supported by Israel, Norway, Ukraine, Morocco, and South Korea. China withdrew from early backing after it decided to develop its own system, Beidou/Compass.

The project was not without controversy—mainly in the form of stern opposition from the US government. Galileo has always been promoted as a *civilian* service; in contrast, GPS (US), Glonass (Russia) and Beidou/Compass (China) primarily have a *military* purpose. Even though the US had started to remove the "selective availability" from GPS—a function that enabled it to reduce the accuracy of the service at a time of conflict—the EU was not happy with the safety implications of that. (In contrast, the Galileo service includes a guarantee of accuracy that makes it more suitable for use in commercial aviation.) The situation made the Pentagon so angry that, according to one European defense official, it threatened to shoot down Galileo satellites if it thought the system was being used by enemies of the US in conflict. The threat was ignored.

The system is based on twenty-seven operational satellites plus three spares. The basic service, accurate to one meter, will be available free of charge throughout the world. A subscription service is accurate to one centimeter. Uniquely, Galileo incorporates an important feature of interest to seafarers: a search-and-rescue facility will not only detect signals from distress beacons; it will also be able send messages back indicating the status of the rescue operation. New secondary features are expected; to keep up with developments visit the European Space Agency website: www.esa.int. See GLOBAL POSITIONING SYSTEM for a fuller discussion of satellite navigation at sea.

galley

the cooking place on a boat or ship.

The world seems to be divided up into two kinds of LIVE-ABOARD; those who believe that their home/boat should have a well-appointed galley and those who believe that you need a single-burner stove and the provisions can be dumped in the bilges. The latter philosophy is based on the conclusion that cooking while under way is so damned difficult that you are only ever likely to take the minimalist approach of throwing everything (including boil-in-the-bag ration packs) into one pot. This divide is not necessarily along male/female lines because I am in the former category, based on a conclusion I have reached that live-aboards rarely spend more than 20 percent of their time making passage. This means of course that 80 percent of their time is spent in an anchorage or a marina.

The last sailing boat I lived on had roughly three-quarters of its main saloon dedicated to the storage, production, and consumption of food. Forward portside was the galley with a full-specification gas stove (four burners, grill, and oven), two too-small sinks, and lots of stowage lockers over and under. Opposite, starboard-side was a big built-in, top-loading refrigerator powered by a 24-volt Adler-Barbour Super-Cold Machine, again with storage lockers above and below. Aft starboard-side was our beautiful table and bench seats. A very satisfying feature of the table was a "secret" booze compartment that would be known only to customs officers. What we actually stored in there were toilets rolls and kitchen rolls. But, between each paper roll we cunningly placed a bottle of wine or liquor to stop the rolls chafing against each other. We also had a baby hammock in which to keep vegetables and fruit. The space below and behind the bench seat was used for non-food items such as spare parts and bo'sun's tools. (For peace of mind I should mention that the other 25 percent of the saloon was occupied by a handsomely fitted-out navigation station.)

The consequence of all this was that we had ample space for provisions while under way and, while on a mooring, we could entertain handsomely. I have never sailed with a family, but the need to keep morale up as well as calorie intake in such a situation might be the clinching argument.

gallows

a frame across the STERN of a vessel used as a cradle for the end of a MAINSAIL BOOM once the mainsail has been lowered and the TOPPING LIFT slackened off; see also YARDARM.

gasket

a short length of rope fixed to a sail and used to secure that sail when FURLED or REEFED.

Genoa ("genny")

a large FORESAIL, the foot of which makes contact with the FOREDECK, extends abaft the SHROUDS, and seriously interferes with the view forward on the side it is set. Gennies come in a variety of sizes but are now most commonly fitted to ROLLER-REEFING gear thus becoming an all-purpose foresail.

genset

short for "generator set"; a petrol (gasoline) or diesel-powered electric generator. Portable 50cc gensets are very useful things to have on boats; they will drive power tools and charge batteries even when a long way from a berth.

get under way

to leave, depart, set off.

GHA

see Greenwich Hour Angle.

giants of Patagonia

a myth concerning the size of some people native to Patagonia.

(Refers to the adventures of Commander John Byron. See also MARINES, MIDSHIPMAN, MUTINY, SHIP FEVER and SQUADRON.)

In the 16th, 17th, and 18th centuries, long before the invention of UFOs, a popular obsession in London coffee shops was with the mysterious "Giants of Patagonia." Patagonia is the region at the southernmost end of South America that spreads across Chile and Argentina and embraces part of the Andes, the Rio Colorado, Tierra del Fuego, terrible weather, and a pretty bleak and hostile coastline.

The story of the giants was all the fault of Ferdinand Magellan. The renowned Portuguese explorer was the first to sail from the Atlantic into the Pacific, which he did through the Strait of Magellan. (Now you know how it got its name.) His flotilla of five ships led by the *Trinidad* was funded by the Spanish Crown and was tasked to find a western route to the Spice Islands, the Maluku Islands in present-day Indonesia. Of the 260 men who left Seville in Spain, only eighteen made it home. Magellan himself was killed in battle in the Philippines. This may seem like a dreadful toll for a few barrels of nutmeg, but the Spice Islands were the only source of what was then prized as a valuable commodity.

Stop thinking of Commander Magellan as a contemporary of Commander Byron. Magellan's fleet set off in 1519; that's 245 years before Foul-weather Jack hauled anchor. But one of Magellan's officers left a legacy.

Antonio Pigafetta, an Italian like Columbus, was among the eighteen survivors in 1522. A scholar and explorer from the city-state of Venice, he worked on the voyage as the commander's secretary and assistant. Throughout, he kept a detailed journal of the first-ever circumnavigation. This is what he wrote about their approach to Patagonia:

> Leaving that place, we finally reached 49 and one-half degrees toward the Antarctic Pole. [Somewhere on the coast of the Argentine province of Santa Cruz, perhaps a little south of Peninsula San Julian.] As it was winter, the ships entered a safe port to winter. We passed two months in that place without seeing anyone. One day we suddenly saw a naked man of giant stature on the shore of the port, dancing, singing, and throwing dust on his head. The captain-general sent one

> of our men to the giant so that he might perform the same actions as a sign of peace. Having done that, the man led the giant to an islet where the captain-general was waiting. When the giant was in the captain-general's and our presence he marvelled greatly, and made signs with one finger raised upward, believing that we had come from the sky. He was so tall that we reached only to his waist, and he was well proportioned. His face was large and painted red all over, while about his eyes he was painted yellow; and he had two hearts painted on the middle of his cheeks. His scanty hair was painted white. He was dressed in the skins of animals skilfully sewn together. That animal has a head and ears as large as those of a mule, a neck and body like those of a camel, the legs of a deer, and the tail of a horse, like which it neighs, and that land has very many of them.
>
> *The first voyage round the world by Magellan* by Lord Stanley of Alderley, 1874. This includes a translation of Pigafetta's journal.

Never mind the llama, Antonio, what about the *giant*? "He was so tall that we reached only to his waist" sounds remarkable. What was the average height of Europeans in the early 16th century? Nobody seemed to be gathering statistics then, but if it was, say, 5ft 6in (1.68m) that would make the Patagonian native 11ft tall (3.36m). And that would make him a giant by any norm. Or had the explorers been at the rum? Why did they see only one Indian? Could he have been standing on a rock? When Pigafetta's journal was published in Europe it caused a sensation. Giants found in Patagonia! Had huge feet!

More than sixty years went by before anyone else reported on the mysterious giants. An English explorer and privateer called Thomas Cavendish set sail for Cape Horn in 1586. Only twenty-six years old, he commanded three ships: the *Desire* (eighteen cannon); the *Content* (ten cannon); and the *Hugh Gallant*. Four months later they were off the coast of Patagonia. This is Cavendish's account of what happened:

> Sailing from Cape Frio, in the Brasils, they fell in upon the coast of America, in 47 d. 20 m. North latitude. [Unless they'd fallen upon the coast of Newfoundland, he means *South* latitude, of course.] They proceeded to Port Desire, in latitude 50. Here the Savages wounded two of the company with their arrows, which are made of cane, headed with flints. A wild and rude sort of creatures they were; and, as it seemed, of a gigantic race, the measure of one of their feet being 18 inches [0.46m] in length, which, reckoning by the usual proportion, will give about 7 feet and an half [2.29m] for their stature.
>
> Thomas Cavendish in J. and J. A. Venn, *Alumni Cantabrigienses*, 1922–1958

I'm not sure about that feet-to-height ratio; that's a measure usually heard about another part of the anatomy. In any case, the Giants of Patagonia seem to have shrunk by three feet. It was time for someone to start measuring these people. The French naturalist Charles Debrosse gave it a try in 1756, not long before John Byron started his voyage.

> The coast of Port Desire is inhabited by giants fifteen to sixteen palms high. [Generally recognized to be the width of the palm of a hand: 3 inches or 7.62cm. However, that would make the giants 3ft 9in (114.3cm) tall and not giants at all.] I have myself measured the footprint of one of them on the riverbank, which was four times longer than one of ours. I have also measured the corpses of two men recently buried by the river, which were fourteen spans long. [Changing units of measure in the middle of a narrative is neither helpful nor scientific.] Three of our men, who were later taken by the Spanish on the coast of Brazil, assured me that one day on the other side of the coast they had to sail out to sea because the giants started throwing great blocks of stone of astonishing size from the beach right at their boat. In Brazil I saw one of these giants which Alonso Díaz had captured at Port Saint Julien: he was just a boy but was already thirteen spans tall. These people go about naked and have long hair; the one I saw in Brazil was healthy-looking and well-proportioned for his height. I can say nothing about his habits, not having spent any time with him, but the Portuguese tell me that he is no better than the other cannibals along the coast of La Plata.
>
> Charles Debrosse: *Histoire des navigations aux terres australes, contenant ce que l'on sait des moeurs et des productions des contrées découvertes jusqu'à ce jour*, 1756. Note the reference to the "land down-under"; reputedly Captain James Cook found this book useful when trying to find Australia.

Giant *cannibals* found in Patagonia! The story gets more tabloid as it goes along.

When Captain James Cook landed in Patagonia he decided to resolve the matter by capturing one of the giants. Nobody onboard the *Discovery* wanted to share a hammock with the newcomer so he was tied to the mast. Quite rightly the guest resented this, broke free of the ropes, bounded over the side and swam ashore. But how big was he?

Cook recorded in his log that he, the captain, was 6ft 3in (1.90m) tall and could stand under the arm of a Patagonian Giant. I assume that he was guessing there; would anyone voluntarily stand under the arm of a giant cannibal?

At least we can rely on the redoubtable John Byron to solve the dispute once and for all. He described his trip ashore in Patagonia as

> putting an end to the dispute, which for two centuries and a half has subsisted between geographers, in relation to the reality of there being a nation of people of such an amazing stature, of which the concurrent testimony of all on board the *Dolphin* and *Tamer* can now leave no room for doubt.
>
> *An Account of a Voyage Around the World in the Years MDCCLXIV, MDCCLXV, and MDCCLXVI by the Honourable Commodore Byron in His Majesty's Ship the Dolphin*, 1767

When Byron went ashore he took with him beads and ribbons as gifts, peace offerings. After all, it cannot be easy to measure a giant cannibal if he is shooting arrows at your chest. The gifts were handed out

> giving to each of them some, as far as they went. The method he made use of to facilitate the distribution of them, was by making the Indians sit down on the ground, that he might put the strings of beads &c. round their necks; and such was their extraordinary size, that in this situation they were almost as high as the Commodore when standing.

The frontispiece to Byron's book shows an English sailor giving a biscuit to one of the Patagonian women. If the illustration is accurate, and the sailor is about 5ft 6in (1.68m), then she is very big indeed, maybe 7ft tall (2.13m). But was the artist there to record the encounter? Or was the drawing made to spice up the book and keep the publisher happy?

Without doubt, some men of the region are tall enough to get scholarships to play college football but, despite the numerous accounts, no surviving *Giant* Patagonians have been found, nor has any archaeological evidence ever been recorded. The dispatches of the explorers did, however, give Jonathan Swift a great idea for a satirical novel.

"gild the gingerbread," to

the gingerbread was the decorative relief work on the high stern of TALL SHIPS to which was added gold leaf; now used to mean "over-egging the pudding."

gill-guy, a

see GADGET.

gimbals

two metal rings, one set inside the other and with hinges offset by 90°; used to enable a cooker (STOVE), COMPASS, or lamp to settle level, independent of the motion of the vessel. Keeps the chili con carne in the pot during rough weather and makes it totally unnecessary to wear your oilskins while cooking. . . . (The author takes no responsibility for the consequences of any reader taking the previous remark seriously. Especially men wearing shorts.)

give way

yield to vessels that have right of way; see GIVE-WAY VESSEL.

give-way vessel

in a right-of-way situation, the vessel that must give way to the vessel that is "in the right" is the "give-way vessel" and needs to respond in a manner specified by COLREGS (US: NAVRULES).

International Rules 14–16:

> **International Rule 14 Steering and Sailing Rules**
>
> Head-On Situation
>
> (a) When two power-driven vessels are meeting on reciprocal or nearly reciprocal courses so as to involve risk of collision each shall alter her course to starboard so that each shall pass on the port side of the other.
> (b) Such a situation shall be deemed to exist when a vessel sees the other ahead or nearly ahead and by night she could see the masthead lights of the other in a line or nearly in a line and/or both sidelights and by day she observes the corresponding aspect of the other vessel.
> (c) When a vessel is in any doubt as to whether such a situation exists she shall assume that it does exist and act accordingly.
>
> **International Rule 15 Steering and Sailing Rules**
>
> Crossing Situation
>
> When two power-driven vessels are crossing so as to involve risk of collision, the vessel which has the other on her own starboard side shall keep out of the way and shall, if the circumstances of the case admit, avoid crossing ahead of the other vessel.
>
> **International Rule 16 Steering and Sailing Rules**
>
> Action by Give-way Vessel
>
> Every vessel which is directed to keep out of the way of another vessel shall, so far as possible, take early and substantial action to keep well clear.

A helmsman needs to be *instinctively* familiar with these rules and cannot afford to dive down the companionway in search of the NavRules whenever there is another vessel in sight.

Global Positioning System (GPS)

a system of twenty-four satellites that continually transmit (on the same frequency) identification and timing information which, when heard by console-mounted or handheld receivers, can be used to deduce an accurate position anywhere on the planet.

The satellites are not in geo-stationary orbits and, therefore, you can never be certain which of a maximum of twelve satellites will be overhead at any one time. But in any case, that job is undertaken by the receiver which, when switched on, will put on a little show selecting a batch of satellites on which it will take BEARINGS. These will then be combined to produce a latitude and longitude for your location. (More on functionality later.)

GPS was developed by the US Department of Defense for military purposes at a cost of $26 billion. The first satellite was launched in 1978, the last of the twenty-four in 1994. The specification of the system was published so that commercial companies could manufacture receivers for the civilian marketplace. The seagoing community was very quick on the uptake, aviators the slowest. Today, it is possible to buy a quite sophisticated handheld receiver for as little as $150 (£100) and the use of the service is free. That has to be the bargain of the millennium.

The only drawback has been "SELECTIVE AVAILABILITY." The Pentagon was concerned that, at times of conflict, the enemy would be able to use the system to obtain advantage over US forces. Selective availability (SA) randomly fuzzed-up the signal so that the derived position for civilian users could not be guaranteed to be more accurate than 100m.

An alternative soon became available when the Russians announced GLONASS. Glonass had no SA and was accurate to 10m. Sadly though, receivers were scarcer than hen's teeth and expensive. The University of Leeds in Britain did some research that showed that Glonass and GPS combined produced by far the best results and, of course, if either system went out of commission, the most reliable. This was an even more expensive solution. Glonass is mostly used today by merchant ships as a backup to GPS.

The next development came when the White House announced that President Bill Clinton, as Commander in Chief, had ordered the Pentagon to switch off SELECTIVE AVAILABILITY. This came as a huge surprise to BOAT/

US and the rest of the yachting community, which had been lobbying to get it turned off for years. Had all of America's enemies disappeared overnight? The answer was much more intriguing than that. For some years, the US Coast Guard has been offering a DIFFERENTIAL GPS (DGPS) service. The charter of the USCG requires it to do everything possible to ensure the safety of people at sea and improving the accuracy of GPS fell within that objective. The bottom line of all this was that the USCG was spending US taxpayers' hard-earned dollars effectively to remove the selective availability put into GPS by the US Department of Defense. Common sense prevailed and SA came out. (It has been claimed that the Pentagon could put it back at any time it feels it necessary but the current generation of satellites, allegedly, does not include an SA function.)

The most recent advance has come from the European Union and the European Space Agency. In 2002 they announced GALILEO, a system using thirty satellites orbiting at 15,000 miles and providing a basic 1m accuracy, 1cm on the subscription service. One-meter accuracy doesn't just tell you where you are in the world; it tells you where the receiver is on the boat. Galileo is due to be fully operational by 2018 at a measly cost of $4 billion.

Even if you are reluctant to go totally hi-tech on your boat, a combination of a simple GPS receiver plus old-fashioned paper charts and drawing instruments will get you ninety percent of the benefits. Add-on functionality found in even the most basic receiver includes WAYPOINTS and CROSS-TRACK ERROR, both of which have changed the way in which we navigate and, indeed, manage our navigation.

As a piece of digital electronics, GPS receivers can interface (using an *NMEA* standard) with other equipment on the vessel, including laptop computers, digital PLOTTERS, radar (to display waypoints), and AUTO-PILOTS. I have had great fun with these and laptops with digital charts on CD-Roms are a very convenient way of selecting waypoints and combining them into lists that can then, after checking, be transferred into the GPS receiver without the risk of transcription errors in the latitude and longitude. But, once the lines are aboard and the fenders stowed, it is GPS and user-friendly paper charts.

If you see yourself as a bit of a dinosaur, consider this: GPS technology would not be possible without the computer. If you plan to stick to sextant-based CELESTIAL NAVIGATION, bear in mind that the reduction tables you use are calculated and produced using computers. Indeed, the very first computer—Babbage's mechanical Analytical Engine—was designed with the specific purpose of producing accurate navigation tables. (Nineteenth-century empire building was won by the country with the best navigation tables—the British.) Finally, the ASTROLABE used to be considered a piece of high technology, once upon a time.

Footnote: Satellite navigation is, so far, the only practical application of Einstein's Theory of Relativity.

Glonass

a Russian satellite-based navigation system; see GLOBAL POSITIONING SYSTEM.

GMT

Greenwich Mean Time; the time at the Greenwich MERIDIAN, the 000° line of longitude. Also known as Universal Time (UT) or even Universal Time Coordinated (UTC) after sidereal corrections. The military call it Zulu Time. See also LOCAL TIME.

going about

the process of turning a sailing boat from one tack to another so that the wind comes over the opposite side of the hull; see TACKING.

going to weather

to sail toward the direction of the wind and/or seas.

gooseneck

a fitting used to attach a SPAR to a mast while permitting the spar unrestricted horizontal and vertical movement.

goosewinged

going DOWNWIND with the MAINSAIL and FORESAIL set at opposite sides of the mast; while the main boom will hold out the mainsail, a WHISKER POLE will be needed to help the jib catch the best of the wind. This unstable kind of arrangement, prone to broaching in heavy seas, is forced upon those with BERMUDAN RIGS because of its design tendency toward windward efficiency.

"go overboard about something"

to overreact; probably originates from the constant risk in boat handling of going a little too far in reaching for (or recovering) something and ending up in the BRINEY.

GPS

see GLOBAL POSITIONING SYSTEM.

GPS Plotter

a navigation instrument that combines a GPS receiver with the ability to display digital charts on a screen; information such as current position, WAYPOINTS, and COURSE MADE GOOD can be superimposed on the chart display.

grab bag

an emergency ship's survival kit; a supply of rations and essential equipment to be taken on the liferaft in case of having to abandon ship. Survival packs placed inside inflatable liferafts are only adequate for a few days, depending on how many crew are onboard. Passage makers are strongly advised to supplement this with additional provisions kept in a watertight container secured on DECK. The checklist I recently used in assembling a grab bag appears on the facing page. Items marked with an asterisk (*) are essential.

The above kit fitted into two watertight containers bought at a chandler's and that were lashed to the pushpit close to the dinghy davits. The FLARES were in an orange watertight box stowed in a cockpit locker. Drinking WATER will need to be kept in separate containers. I always recommend that each crew member has a personal survival kit and is responsible for having it with them in a survival situation. An effective way of encouraging crew to keep such personal kits to hand is to suggest they keep items such as passports, driver's license, cash, credit cards, and traveler's checks in them. Place individual items in waterproof Ziploc® bags and stow the whole thing in a sturdy hiker's belt bag. Separate bags are not allowed if being winched off a vessel by helicopter.

grapnel

a small ANCHOR with three or four arms; some of these are available in a form where the FLUKES fold flat and are suitable for a DINGHY or INFLATABLE.

gray water (US)

a euphemism for what is lurking near the HEAD (UK: heads) in the HOLDING TANK under the CABIN SOLE.

great circle

any line of circumference around the world; therefore, all lines of longitude (going through both poles, of course) and the equator are "great circles."

Grab Bag		
*	"Space blankets"	These are waterproof, heat-retaining, heat-reflecting and compact. Inexpensive—take one per person. Can be used ashore to make shelters.
*	Compass	This can be a spare nautical hand-bearing compass or one of the type used by trekkers.
*	Flashlight	This will enable you to see in the dark but will also serve as a signaling device if it is powerful enough.
	Diary	A pocket diary will help you keep track of time and maintain a log.
	GPS receiver	This is a good place to keep the backup for the console-mounted receiver.
	Sextant	A cheap plastic sextant can help you get a latitude but it will take up a lot of space. A protractor and a piece of string plus a weight will be nearly as good and can be used to draw lines on a chart. (Don't forget a few pencil stubs.) To reduce a sun-shot for a longitude will of course require you to have an almanac and reduction tables onboard.
*	Charts	Charts of the area in which you are sailing. This can be expensive; a low-cost option is to print digital charts from your computer and seal in clear plastic.
*	Strong sharp knife	A small grinding stone will help to keep it sharp.
*	Wristwatch	Even the cheapest modern quartz watches are more accurate than Harrison's best chronometer.
	An axe and/or machete	In case you fetch up at a remote beach.
	Water maker	A hand-operated reverse-osmosis device for converting sea water to drinking water. Expensive, but could save your life if you are adrift for a long time.
*	First aid kit	The type designed to be kept in a car is good, but include extra items like sunblock and aspirin.
*	Food	Canned, dehydrated, or "boil-in-the-bag." Military ration-pack are good, as are long-life high-energy foods such as Kendal Mint Cake.
	Small cooker	A "hexi-burner" with plenty of spare hexi blocks and waterproof matches. You are unlikely to use this in an inflatable, but you might in a hard dinghy and you certainly will on a beach.
*	A fishing kit	Make up your own with plenty of spare hooks, line, and a few weights.
*	Cord	Para-cord is best.
	VHF Radio	If you have a spare handheld radio set this is a good place to keep it.
*	Distress signals	We had so many flares and rockets that we kept them in a separate watertight container on DECK.
*	Batteries	Spare batteries for everything that might need them.

great circle routes

a strategy for long-distance sailing (many hundreds or even thousands of NAUTICAL MILES) based on following a great circle that intersects both the departure and destination points.

The logic for this is simple: it is the shortest distance. This might not be immediately apparent when seeing such a course plotted on a chart using the MERCATOR PROJECTION. In order for maps and charts to be flat (so they can be printed on paper) the image has to be distorted; higher latitudes are made proportionately wider than lower latitudes. If you are having trouble following this, here is an experiment to try. Find a map that shows the North Atlantic on one page (an atlas will do) and, using the scale, measure the distance between Land's End in the UK and New York (or San Francisco to Hawaii if you wish). It doesn't matter whether you use statute miles or NAUTICAL MILES—so long as you use the same for units in the next stage. Now you need a globe. Place one end of a piece of string at Land's End and extend it by the shortest distance across the Atlantic until it reaches New York. Now measure the length of string needed and compare it with the distance measured off the map. The string measurement (the Great Circle Route) will be much shorter.

If you used a conventional chart to plot such a course and you followed it diligently, your COURSE made good would be a long curve. Following a straight line drawn on a Great Circle chart would mean that you were following the shorter distance made by the string on the globe. In practice, a compromise is made whereby a series of WAYPOINTS is used to represent the Great Circle route. If you have ever wondered why Transatlantic races seem to go so far north, look again at the route traced by your piece of string across the globe; it may be against the wind and close to the icebergs, but it is the shortest way. Sometimes competitors will gamble on taking a more southerly route to catch a more favorable wind, one that will compensate for the longer distance traveled.

If you are only traveling a few hundred miles or less, you can safely ignore all this.

See PLANE SAILING.

green

color usually associated with STARBOARD.

gripe, to

when a yacht has excessive WEATHER HELM and keeps driving head-to-wind, she is said to be "griping."

gripes

1. ropes used to secure a ship's boats (tender, lifeboat, or whatever) to the DAVITS.
2. an intestinal spasm.

grog, groggy

a concoction of rum and water. A traditional drink for Royal Navy seamen, issued onboard ship whenever there was the slightest excuse (e.g., the beer running out). The rum-and-water solution was introduced in 1740 by Admiral Edward Vernon. Vernon's nickname was "Old Grogram" after the coarse fabric (grogram) cloak he wore. Hence, "grog" and, more commonly today as an adjective, "groggy," to mean that someone is close to being THREE SHEETS TO THE WIND.

See also PUNCH.

grommet

an eye made in the corner or edge of a sail through which a line can be passed; also a ring made of rope for use as part of the rigging.

Gros Islet

pronounced "Gross-ee-LAY." Gros Islet is a fishing village that you pass to port as you enter Rodney Bay Lagoon, heading into St. Lucia's main, if not only, marina. Once tied up you will probably be directed toward the fleshpots of the island's main tourist area to the south, between Rodney Bay and the capital, Castries. But if your palette fades, and you are of the right turn of mind, catch Gros Islet on a Friday night. It's recommended to go in a group or failing that, get a friendly taxi driver to drive you around there—and stay with you. Shortly after nightfall, the place will be heaving with locals. Bob Marley will be modulating the walls of the pastel-painted wooden houses and the aroma of exotic cheroots will be hanging in the still air.

A couple of blocks up from the beach you will find Mama's Bar in the front room of Mama's House. Now I'd like to tell you that Mama will welcome you with a huge smile and offer you fancy expensive cocktails with umbrellas and plastic sticks that poke your eye out. Not so. Regardless of any frivolity that might be going on around her, Mama takes a serious view of the human condition. Tell her you want "rum" and she'll reach under the improvised counter and produce a half bottle of bootleg dark rum. Then say "mango, please" and a carton of juice will appear. Hand over a few dollars and you will get a couple of plastic beakers and you can mix your own Rot-gut Rum Punch. (Forget about ice; unless Mama is now backed by the Hilton Corporation, she doesn't have a refrigerator.) But the main thing is to enjoy the company; fishermen, local wannabe gangsters, off-duty cops, and the more adventurous tourists all gravitate to Gros Islet at weekends. You can even dance in the streets.

ground tackle

the ANCHOR, the anchor CHAIN, and anything else that might go to the bottom of the sea with them.

guardrail

see LIFELINES.

gudgeon

a metal plate fixed to the TRANSOM of a dinghy or rowing boat; the purpose of the gudgeon is simply to provide holes into which the PINTLES of a RUDDER may be slotted.

gunkhole (US)

to "gunkhole" means to launch the TENDER and explore coves, creeks, marshes, or other shallow areas around a shore.

guns

see FIREARMS.

gunter rig

a small-craft rig that resembles a GAFF RIG but in which the MAINSAIL GAFF lies vertical to (and against) the mast.

gunwhale (pron. "gunn'l")

the top edge of a boat's hull or a rail that circumnavigates the DECK.

guy

any line used to control or steady a SPAR or BOOM.

gybe

see JIBE.

Gypsy

1. a "gypsy" is a wheel on a WINDLASS with indentations into which the links of the ANCHOR CHAIN engage.
2. see OCEAN GYPSIES.

gyrocompass

a direction-finding compass that incorporates a gyroscope, a heavy metal wheel that rotates inside a frame; when the compass is started, it is orientated to north and keeps that orientation as the frame (and the vessel) move around it. When it was introduced in the early 20th century, the electrical gyrocompass represented the first move away from the magnetic compass for use in metal ships. And now of course the gyrocompass is being replaced by even more exotic technology including ring lasers and digital flux gates.

H

halyard

a line used to haul a sail up into its working position; in practice, there would be a separate halyard for the MAINSAIL, the JIB (or GENOA), the SPINNAKER, and the MIZZEN sails; in modern yachts, halyards are often routed inside the relevant mast. The word probably derives from "haul" and "YARD," a "haul-yard." A halyard is also used to raise a flag or ensign.

hammock

a suspended bed made of CANVAS or netted line. It was invented by the Arawakan people of the Caribbean who, presumably, slung them between two COCONUT-free palm trees. The Spanish saw them first and called them *hamaca*.

hand

to "hand" is to pull or haul on something; for example, after releasing a HALYARD, the sail is then "handed" to pull it all the way down to the boom.

hang off

to secure one rope to another so that it will take the strain while something is done with the end of the first; a technique commonly employed to clear jams on WINCHES.

hank

a metal or (on a dinghy) plastic hook used to secure the LUFF of a FORESAIL to a STAY.

hard, the

a firm sloping FORESHORE, often artificial, that can be used as a landing place; or otherwise just a place where trees grow.

hard chine

a sharp intersection between the HULL's side and bottom.

harden in

see HARDEN UP.

harden up

to LUFF UP closer to the wind; the expression derives from the need to pull in and tighten the main and fore-sheets.

hard over

the nautical equivalent of "full lock" on the steering of a car; the WHEEL is turned or the TILLER is pushed/pulled over as far as possible.

hard tack

a seriously solid biscuit used as a substitute for BREAD on Royal Navy ships until 1855.

> It was an automatic gesture of every seaman to tap his hard-tack on the table before eating it to knock out any maggots and weevils. One nineteenth-century seaman, Dalrymple Hay, described hard-tack so riddled with them that the only way to make it palatable was to lay the entrails of a freshly-caught fish on top of a mound of it. The maggots and weevils came out of the biscuit into the entrails, which were then thrown away.
>
> *The Custom of the Sea* by Neal Hanson, 1999

Note the use of *fresh* fish entrails, not any old entrails that might have been lying around for a few weeks. If you think Dalrymple Hay might have been going over the top, this is what Donald Thomas wrote about the matter:

> Corned beef and biscuits, wine and water made up much of the diet in ships of the Atlantic squadrons [of the Royal Navy]. The beef, reported young Bernard Coleridge, had been ten or eleven years in the corn. The biscuits felt like cool calves' foot jelly or blancmange when swallowed because of the number of maggots in them.
>
> *Cochrane: Britannia's Sea Wolf* by Donald Thomas, 2001

Never miss out on a chance to increase your protein intake. See BREAD.

hardtop

a fiberglass BIMINI or roof over the cockpit and helm area.

hatch

an opening in a DECK to allow in and out people, cargo, sunlight, provisions, sails, ventilating air, and, if the hatch is open at the wrong time, a great big gush of sea water. Hence the need for hatch covers.

haul

to pull something.

haul around

to change course in such a way that a sailing boat RUNNING before the wind turns onto a REACH.

hawse hole

holes in the topsides at the bow through which the ANCHOR cable (RODE) or shore lines can run.

hawser

a line used for towing.

hazardous marine creatures

If I listed all the dangerous critters in the sea it would (a) take up the whole book and (b) frighten you away from ever putting a toe into the BRINEY—and I would not want to do that. So, just to frighten you a little bit, here is a roundup of the usual suspects.

<table>
<tr><th>CREATURE</th><th>HAZARD</th><th>TREATMENT</th></tr>
<tr><td>Cone Mollusks

Marine snail with attitude.

Found in the Indian Ocean through to the western Pacific.

Very pretty. Not very friendly.</td><td>Has a tooth-like spike on the end of its tongue. While you are admiring it, it shoots its tongue out and stabs your hand. The venom numbs your flesh and causes paralysis. This spreads to your lungs and you die.</td><td>There isn't any treatment. You could try praying that someone on board knows Cardiopulmonary Resuscitation (CPR).</td></tr>
<tr><td>Jellyfish and variations on the theme (e.g., the Portuguese Man-o'-War and sea wasps).

These occur in the tropical and subtropical latitudes of the Atlantic, Pacific, and Indian Oceans.

The Portuguese Man-o'-War is known to follow the Gulf Stream.</td><td>Five species of jellyfish can cause death. They inject a venom that causes paralysis of the respiratory system.

The villain of the piece, though, is probably a sea wasp called Chironex fleckeri.

This is usually about 20cm in diameter with 3m long tentacles. The sting can kill you within 4 minutes—which makes the blue-ringed octopus look like a wimp.</td><td>You might have a chance here, because the venom is relatively short-lived.

Dig out as much of the tentacles as you can. (If done in the cockpit you can wash away the blood easier.)

For Chironex fleckeri only, wash out the wounds with lots of vinegar.

The Australians have an antivenin for this one, so let's hope you get stung there.

With the Atlantic jellyfish Chrysaora, use a 50% solution of baking soda in water.

Then go for CPR and keep trying. This is one you can win.</td></tr>
<tr><td colspan="3">Do NOT use any form of alcohol on any jellyfish stings: it stimulates release of venom</td></tr>
<tr><td>Octopuses

Eight-legged shell-less mollusk—the stuff of Hollywood Legend and the Greek dinner plate.

Some species known to grow to 5.5m (22ft) in length.

Never seen one of these on a Greek plate, though.</td><td>Never mind the suckers—under those arms is a powerful beak.

Once the octopus gets its beak into your flesh it tends to slaver with excitement. The saliva is toxic—very!

In particular, watch out for the little 10cm-long blue-ringed octopuses of Indo-Pacific waters. One bite and you are dead within 20 minutes.</td><td>No treatment. No antivenin. No chance.

The octopus is totally deaf, so hurling colorful insults at it won't do any good.</td></tr>
</table>

Creature	**Hazard**	**Treatment**
Snakes Sea snakes are found in warm, calm coastal waters—the sort you like to swim in.	Snakes fall into the same category of risk as sharks. Although they can be very venomous, snakes are not very aggressive and are more likely to be frightened of you. Just don't get them cornered or try to sell them any insurance.	If you get bitten, don't panic. Certainly don't start stabbing at the wounds and trying to suck the venom out—there's probably none in there. Current advice seems to be to get the patient to relax and keep the affected limb as still as possible.
Sharks Have big teeth to bite you with. Found all over the world.	Unless you go looking for trouble (e.g., putting on a black wet suit and impersonating a tasty seal) you are more likely to win a lottery jackpot than be mauled by a shark. If you are fishing off your boat do not be surprised if sharks come by to check out the action. Do stay in the boat, though.	Twelve hours of emergency surgery? The Australians claim to have developed an anti-shark device that can be worn while swimming, but it may be some time before it reaches the market at an affordable price (proportional to the risk).
Stingrays A fish common to tropical waters, but prolific enough to be found as far north as the Mediterranean Sea, North American waters, and even off Scandinavia. Also found in South American rivers. Can grow up to 5m in length and over 300kg._	A very aggressive animal that can whip you with its long tail and inject you with a particularly potent venom. Death can result in some cases.	See the treatment for Stonefish, etc. See also stingrays.
Stonefish, Toadfish, Scorpionfish, and Weever fish		
The **Stonefish** is an ugly brute and, if it had a shoulder, there would be a chip on it. The **Toadfish** is unlikely to win any beauty contests, either.	The dorsal fins of the Stonefish and Scorpionfish comprise a series of sharp venom-loaded spines.	The remedy for these stings is immersion in very hot water—as hot as you can bear without scalding. Even if the pain seems to have subsided, keep at it for up to two hours. There is an antivenin for Stonefish.

Creature	Hazard	Treatment
Stonefish, Toadfish, and **Weever fish** lurk half under the seabed in shallow water. **The Scorpionfish** (also known as the Zebrafish), is far easier to spot, so you are unlikely to tread on it. **Weever fish** can be found as far north as the British Isles.	The pain from these can be excruciating. Death is rare but usually results from heart failure.	If you are likely to enter shallow water where any of these characters operate, wear something on your feet.

In all the above cases you should seek the earliest medical attention. You are most likely to be near the coast, probably just anchored off it, but that doesn't mean you are anywhere near civilization. I would certainly make a PAN-PAN MEDICO call to get expert medical advice. I would also get advice on the possibility of an early evacuation of the victim.

See also CIGUATERA about fish that are dangerous to eat.

head

the top of any part of the RIGGING; the MASTS and SAILS. Can also mean the BOW. If you are looking for the loo (UK), see HEADS.

headfoil

a metal extrusion fitted to a FORESTAY and used to hold the BOLT ROPE on the LUFF of a JIB or GENOA sail; a key component in ROLLER REEFING gear.

head-on situation

you are in a power-driven vessel when you see another powered vessel heading straight at you: what do you do? COLREGS (US: NAVRULES) International Rule 14 applies:

> (a) When two power-driven vessels are meeting on reciprocal or nearly reciprocal courses so as to involve risk of collision each shall alter her course to starboard so that each shall pass on the port side of the other.

(b) Such a situation shall be deemed to exist when a vessel sees the other ahead or nearly ahead and by night she could see the masthead lights of the other in a line or nearly in a line and/or both sidelights and by day she observes the corresponding aspect of the other vessel.
(c) When a vessel is in any doubt as to whether such a situation exists she shall assume that it does exist and act accordingly.

Of course, if it is a super-tanker that is bearing down on you, you are not going to have to think twice about a clear move to starboard.

headroom

the distance between the CABIN SOLE and the DECKHEAD; usually measured on an unfamiliar boat by the number of times you bang your head.

heads, head (US)

a toilet on a boat; the nadir of plumbing expertise. Most commonly a totally unreliable system of pumps and SEA COCKs that is supposed to enable you to use the heads, pump it out, then pump in clean sea water. There is an alternative system of American origin that employs half a basketball. The head is installed in such a way that the water level in the pan is the same as the sea level outside. After using the facility, the lid is locked down and the basketball fitted to the lid depressed firmly and the contents ejected. Having used both, I think I would rather be depressed by the other system.

Restrictions on the use of these unseemly processes in coastal waters has seen the introduction of regulations (in the US, for example) requiring that yachts be fitted with HOLDING tanks. The output goes from the heads to the tank and from there to a honey-pump (if in harbor) or out to sea (if beyond coastal waters). Supposedly derives from "beakhead," part of the FORECASTLE used by sailors as a toilet. See also FOA, "BUCKET-AND-CHUCK-IT," and THUNDERBOX.

headsail

any sail rigged forward of the foremost mast.

head seas

waves coming from ahead toward the BOW of the boat.

headway, making

"making headway"; beating to windward and making ground, now in the vernacular as "making progress against the odds."

headwind

any damned wind coming from the direction in which you wish to travel.

heave

1. to pull on or to haul a line.
2. to throw a line; see HEAVING LINE.

heave to

just about the only way of stopping a sailing boat—short of ANCHORing or mooring. The conventional way of doing this quickly (e.g., in a man overboard situation) is to turn to windward until the jib (still sheeted to what was the leeward side) backs. Then push the tiller (or turn the wheel) to the (new) leeward side. This should have the effect of bringing the yacht's BOW toward the wind and holding it there. This is worth practicing. The ability to stop a boat by heaving-to can be employed in a number of different situations from storm survival conditions (where it will also be necessary to reduce sail) to the mundane but important need to stand off from an unfamiliar port until daylight. Here is Jack London's hero Chris Farrington battling to survive in a TYPHOON.

> Working as well as he could with his bandaged hand, and with the feeble aid of the Chinese cook, Chris went forward and backed the jib over the weather side. This with the flat mainsail left the schooner hove to.
>
> "God help the boats! It's no gale! It's a typhoon!" the sailing-master shouted to Chris at eleven o'clock. "Too much canvas! Got to get two more reefs into that mainsail, and got to do it right away!" He glanced at the old captain, shivering in oilskins at the binnacle and holding on for dear life.
>
> "There's only you and I, Chris—and the cook; but he's next to worthless!"
>
> In order to make the reef, it was necessary to lower the mainsail, and the removal of this after pressure was bound to make the schooner fall off before the wind and sea because of the forward pressure of the jib.
>
> "Take the wheel!" the sailing-master directed. "And when I give the word, hard up with it! And when she's square before it, steady her! And keep her there! We'll heave to again as soon as I get the reefs in!"
>
> Gripping the kicking spokes, Chris watched him and the reluctant cook go forward into the howling darkness. The *Sophie Sutherland* was plunging into the huge head-seas and wallowing tremendously, the tense steel stays and taut rigging humming like harp-strings to

the wind. A buffeted cry came to his ears, and he felt the schooner's bow paying off of its own accord. The mainsail was down!

"Chris Farrington, Able Seaman" from *Dutch Courage and Other Stories* by Jack London, 1901

heaving line

a line or cord arranged with a weight (or a neatly-tied MONKEY'S FIST) at one end so that it can be accurately thrown a long distance. A heavier line can then be attached and pulled across.

An alternative is the throw-bag or throwing-bag, whereby the end of the line is coiled in a canvas bag. As the bag is thrown, the line feeds out until the bag drops neatly at the feet of the recipient. The sub-text to the system is that, in the unlikely event that you miss, you haul the bag back, filling it with water on the way. Then you throw it again, generally spraying the water all over your companions and the recipient. When this happens, all you can do is offer to buy the first round of drinks.

heel

the angle of a boat from the vertical, caused by the pressure of wind on sail, when sailing. See STIFF.

Helios

a Greek god with a golden boat.

> The Greek sun-god, who rode to his palace in Colchis every night in a golden boat furnished with wings. He is called Hyperion by Homer, and, in later times, Apollo.
>
> *The Brewer Dictionary of Phrase and Fable*, 1870

The Greeks had the right idea, but I wonder if "a golden boat" can have the electrolysis problem? There you are in the cockpit with your feet up and thinking that all is well . . . but meanwhile, below the WATERLINE all that gold is being corroded away.

helm

1. the device by which a boat is steered (ultimately by changing the angle of the RUDDER to the FORE-AND-AFT line of the vessel); may be a TILLER or a WHEEL.
2. the person steering the vessel; these days, of course, it may be a computer, even on a small yacht.
3. the part of a boat where steering and other controls are located.

hitch

a means of making the end of a rope fast to some object (e.g., CLOVE HITCH).

hoist

1. the hoist: the vertical side of the flag next to the mast; also used to refer to its vertical edge.
2. to hoist: to HAUL something upward.

hold, to

when an ANCHOR gets to grips with the sea bottom, it is said to be "holding."

holding tank

storage tank for output from the HEADS (UK) or HEAD (US). Once full, can be emptied at an appropriate pumping-out station or via a through-deck fitting and into the sea if sufficiently offshore to meet local coast guard and environmental requirements (and on pain of a huge fine).

Hoodoo Triangle, The

see BERMUDA TRIANGLE.

hook, the

nickname for the ANCHOR.

horizon, the

the line where the sea appears to meet the sky; it is important that the horizon be visible when measuring the elevation of a CELESTIAL OBJECT.

I was sailing with a chum in the Irish Sea, heading south in his 26ft Trapper 300 from Scotland to the Isle of Man for a serious circumnavigation. We were completely surrounded by . . . nothing. Except horizon. I was navigating and helming, he was trying to repair the rusty socket used to get power to the autopilot. He was also suggesting that we were well off course. I suggested that the problem really lay in his insistence that his tub was totally free of DEVIATION—an improbable circumstance. When I next looked up, a huge container liner was passing across our BOW from starboard to port about 200m away. My friend looked where I was looking and, by the time he had turned back to me, the blood had drained from his face. I maintained our heading and said (coolly, I thought), "We are the STAND-ON VESSEL."

Because the world is big (I suppose) it is a common misconception that the horizon must be a considerable distance away. It isn't. If, for example, you are standing in the cockpit of a small yacht, and your head is 2m (6.6ft) above the surface of the water, the horizon is only 2.9 NAUTICAL MILEs away. The funnels of a large ship will begin to appear when it is about 7nm on the other side of the horizon—in other words, about ten miles away depending on the height of its superstructure. So, if it is making 15 KNOTs, it will be on top of you in 40 minutes. If it is a hovercraft or a large passenger-carrying catamaran, you could be under its BOW in 10 minutes. It is the sort of thing that could ruin your nap. (By the way, I was right about the deviation.)

horse

a rope, piece of wood or, more usually today, a metal track which runs across the DECK (ATHWARTSHIPS) and to which the inboard end of a SHEET is attached; used to enable a better downward pull on the sail, thus flattening it for more effective use when sailing to windward.

hove down

excessively HEELED.

hove to

see HEAVE TO.

huffle

1. a man who temporarily joins the crew of a Thames Sailing Barge when mast and sails have to be lowered to pass under a low bridge.
2. a docker (US: longshoreman) who takes the lines of a ship as she approaches her berth.

hull

the floating body of a boat; excludes the rigging in the case of a sailing boat.

hull speed

the maximum SPEED of a yacht; primarily a function of the length of the hull at the WATERLINE.

hurricane

the name of Carib origin (*huracan*) for a rotating TROPICAL STORM.

Hydrographer of the Navy (UK)

Chief Executive of the UK Hydrographic Office, a defense agency employing a staff of 800 civilians and some Royal Navy personnel. The Office is responsible for meeting national hydrographic needs and provides the hydrographic products and services, both digital and analog, that mariners need to go about their business at sea, safely and effectively. In practical terms, this means charts, pilots, and tide tables. See also ALMANACS.

I

IALA

International Association of Lighthouse Authorities; the international standards body for matters concerning navigation. See BUOYAGE.

immunization

see JABS.

inboard engine

an auxiliary engine installed amidships and as low as possible in the hull; see OUTBOARD ENGINE.

inflatable boat

a small boat that can be folded up for stowage and then inflated for use; air is pumped in using a foot pump, an electric pump, or automatically using a compressed gas cylinder in the case of a liferaft. If the boat has a rigid bottom, it becomes a "RIB"—A RIGID INFLATABLE BOAT.

inflatable pillows

catching sleep where and when you can is always a challenge onboard, especially if you are sailing single-handed. It is sometimes the case that you need to slump onto a pilot berth or a settee in the saloon with your oilies on. Getting your head comfortable is the key to achieving those few precious moments of slumber before the jib needs changing again. If you use pillows, they are going to end up wet and soggy and stay that way until you hit the tropics. A practical alternative that works for me is one of those inflatable headrests sold in airports to airline passengers. It does not matter if you get them wet, and they stop your head from rolling about with the movement of the boat.

As an alternative, take a pair of short rubber boots, stuff the leg of one inside the other and wrap in a sweater or towel. This does not work, of course, if the boots are on your feet.

Inland Rules (US)

Navigation Rules (NAVRULES) applicable only to the inland waters of the United States (great rivers, Great Lakes, etc.).

inoculations

see JABS.

International Code of Signals

a system of signaling managed by the INTERNATIONAL MARITIME ORGANIZATION. See SIGNAL FLAGS.

International Dateline

a line of LONGITUDE 180° east and west of the GREENWICH MERIDIAN at which the onboard date is advanced or retarded by one day for vessels crossing it. In practice, the line diplomatically negotiates its way around various groups of islands so that only one date need appear on local newspapers.

International Maritime Organization

The International Maritime Organization (IMO) is the United Nations' specialized agency responsible for improving maritime safety and preventing pollution from ships. This is how the IMO describes its responsibilities:

> Shipping is perhaps the most international of all the world's great industries—and one of the most dangerous. It has always been recognized that the best way of improving safety at sea is by developing international regulations that are followed by all shipping nations and from the mid-19th century onwards a number of such treaties were adopted.

Several countries proposed that a permanent international body should be established to promote maritime safety more effectively, but it was not until the establishment of the United Nations itself that these hopes were realized.

In 1948 an international conference in Geneva adopted a convention formally establishing IMO. It entered into force in 1958 and the new Organization met for the first time the following year.

Its first task was to adopt a new version of the International Convention for the Safety of Life at Sea (SOLAS), the most important of all treaties dealing with maritime safety. This was achieved in 1960 and IMO then turned its attention to such matters as the facilitation of international maritime traffic, load lines and the carriage of dangerous goods. . . .

Shipping, like all of modern life, has seen many technological innovations and changes. Some of these have presented challenges for the Organization and others, opportunities. The enormous strides made in communications technology, for example, have made it possible for IMO to introduce major improvements into the maritime distress system.

In the 1970s a global search and rescue system was initiated. The 1970s also saw the establishment of the International Maritime Satellite Organization (INMARSAT) which has greatly improved the provision of radio and other messages to ships.

In 1992 a further advance was made when the Global Maritime Distress and Safety System became operative. When it is fully in force in 1999 it will mean that a ship that is in distress anywhere in the world can be virtually guaranteed assistance, even if the ship's crew do not have time to radio for help, as the message will be transmitted automatically.

International Regulations for Preventing Collisions at Sea, 1972

a set of rules published by the INTERNATIONAL MARITIME ORGANIZATION aimed at preventing collisions at sea; no surprise there, then. These regulations have the force of maritime law. The shorthand is COLREGS (UK) or NAVRULES (US).

inverter

a device that converts direct current (DC) electricity (from the 12v batteries) to alternating current (AC) of the type that comes out of the wall sockets at home (usually 115v/60Hz or 220v/50Hz depending on geography). Enables you to drive the television set, the video recorder, the microwave oven, the laptop computer, the electric toothbrush, the satellite phone. . . .

I remember sitting in the engine room of a newly fitted-out Colvin Gazelle trying to figure out what was wrong with the electrical system. There were four batteries: two in parallel providing 12v and two in series providing 24v. (The 24v was needed because of long cable-runs to the windlass and refrigerator.) The 12v arrangement was charged from either the alternator on the engine or from 110v AC shore supply. The 24v rig was charged from, er—What was it being charged from? The shore supply was there all right. But at sea the alternator charged the 12v batteries, which were connected to an *inverter*. This inverter produced 110v AC, which was then used to charge the 24v batteries. Needless to say, so much energy was being absorbed in this arrangement that the 24v system rarely got its full WHACK and the beer got warm. The author Poul Anderson once declared that he "had yet to see any problem, however complicated, which, when looked at in the right way, did not become still more complicated." The solution? Fix a hefty 24v alternator to the engine—expensive, but much more practical.

"Irish Hurricane"

dead calm; no comment.

iron spinnaker, the

sometimes "iron topsail"; the engine, see DIESEL ENGINES.

irons, in

1. a sailing boat is said to be "in irons" when it is *head-to-wind* with the sails not working and unable to move off onto either tack; a consequence of GOING ABOUT with insufficient WAY on.
2. placed in shackles and thrown into the BRIG as a punishment.

isolated danger marks

a system of BUOYAGE used to mark the position of a surface or underwater hazard to shipping: employs two black balls (one above the other) as a TOPMARK and a distinctive color scheme of red and black bands.

J

jabs (US: shots)

just when you thought it was safe to cast off. . . . Take medical advice about the inoculations recommended for the region in which you will be sailing. (See the table on pp. 159–160.) If you are not sure where you will fetch up, it is recommended to get the jabs before you cast off.

Seek professional medical advice. This is a guide only.

jack

old nautical term for a seaman, a "jacktar"; reputedly so-called because they would cake their hair with tar to keep the lice away. An alternative and perhaps more credible explanation is that waterproof clothes were made from tarred canvas—tarpaulin. The English first started to tackle the problem of longitude in the middle of the 17th century. It was the big scientific challenge of the day and some of the top thinkers met periodically at Gresham College to consider various solutions. The following delightful but anonymous poem from 1661 recorded their activities:

> The Colledge will the whole world measure,
> Which most impossible conclude,
> And Navigators make a pleasure
> By finding out the longitude.
> Every Tarpalling shall then with ease
> Sayle any ships to th'Antipodes.

Condition	Effective for up to . . .	Comments
Malaria An intermittent and remittent fever caused by a protozoan parasite that invades the red blood cells and is transmitted by mosquitos in many tropical and subtropical regions.	0 years	There is no immunization against malaria, so don't even ask. See MOSQUITO.
Hepatitis A A viral infection that can be acquired anywhere in the developing world—and 5% of travelers are thought to do just that.	1 year (10 with booster)	Havrix is now the recommended vaccine. It can be a month before it takes effect and it is expensive. Also effective is actually catching Hep A. It provides lifelong immunity for nothing (apart from any hospital bills).
Hepatitis B Can be caught anywhere in the world mainly as a result of unprotected casual sex, the use of shared needles when injecting drugs, or the use of dirty needles and instruments by medical staff.	5 years with booster	Three or four injections are required (over a period of 6 months) with a booster every five years. You have a decision to make here. If you are not into casual sex or sharing syringes full of smack then you might not consider yourself to be at risk. However, should you have an accident or fall ill the vaccine will protect you in an unhygienic hospital or from contaminated blood.
Tetanus	Up to 10 years	You may have been immunized as a child, but ten-yearly boosters are recommended. There is of course an anti-tetanus injection that can be given in the case of accidents where the skin has been broken. This assumes of course that you are anywhere near a hospital that has the stuff when you have the accident. With a longevity of 10 years, this is one I always keep up to date.

Condition	Effective for up to . . .	Comments
Polio	Up to 10 years	You may have been immunized as a child, but ten-yearly boosters are recommended. Polio was thought to have been eradicated, but a new strain of paralytic polio has emerged recently.
Diphtheria Asia, Africa, Central, and South America	Up to 10 years	You may have been immunized as a child, but ten-yearly boosters are recommended.
Tuberculosis (TB) This is on the increase again throughout the world.	Lifelong	Most adults in developed countries are immunized against TB during childhood. Should not need boosting.
Cholera Anywhere where food and water might be contaminated	6 months	This is a wicked jab but, luckily, it's one that you can skip. It's now only recommended for travel to countries where health officials might insist on it. I have known western doctors to fake them.
Rabies The Indian Subcontinent is the high-risk area.	1 year (2–3 with booster)	Requires three injections over a period of six months to one year followed by three-yearly boosters. There are a lot of stray dogs in India, Pakistan, and Sri Lanka, but unless you are planning some overland travel, skip this one, stay on the boat, and hope the dogs can't swim too well.
Typhoid Main risk areas are India and some parts of South America.	3 years	Immunization is now available orally as well as by injection. It involves swallowing 3 capsules over number of days and is more expensive than the jabs. Protection is less than 80% in both cases, so avoid contaminated water and food.
Yellow Fever Prevalent in the tropical latitudes of Africa, Central America, and parts of South America.	10 years	Yellow fever is incurable and deadly, so get the jab if you are likely to enter these areas. Many port health officials will refuse you entry if you do not have a current International Certificate of Vaccination for yellow fever when arriving from an endemic region.

"Tarpalling" is now spelled "tarpaulin" and was a nickname for sailors because they used tar-covered canvas as a waterproof clothing; inevitably abbreviated to "tar."

jack line

see JACKSTAY (2).

jackstaff

a small flagpole fitted to the BOW of a vessel.

The UK's "Union Jack" is really a small version of what is properly called the "Union Flag" intended for flying from a jackstaff of a Royal Navy ship.

jackstay

1. a line that pulls the LUFF of a TOPSAIL close to the mast; also called a "leader."
2. (US: jack line) a safety line that runs along a yacht's DECK from the COCKPIT to its BOW; used to secure a SAFETY HARNESS; probably from "JACK"—an old term for a sailor—and "STAY" for secure. Jackstays are usually made up from plastic-coated wire, but it is remarkable easy to slip on these. A better arrangement is to employ the ultra-strong webbing straps used by mountaineers; they stay flat on the DECK and do not slip so easily.

jam cleat (jamming cleat)

a type of CLEAT that enables a rope to be secured by forcing (jamming) it into a grooved slot; the rope can easily be freed by pulling up out of the slot.

Japan

see EAST.

Japanese encephalitis

see MOSQUITO.

jetsam

debris jettisoned from a ship and floating on the sea (from the Latin *jactāre*, to throw); not quite the same as FLOTSAM.

jetty

a structure that projects out into a body of water (and to which vessels can be secured).

jib

a form of HEADSAIL, secured to the FORESTAY of the MAINMAST, flying free and controlled by SHEETS running to the COCKPIT at the STERN. Not to be confused with a STAYSAIL, which is hanked to an inner stay between the jib and the mainmast. See also GENOA.

jib, "the cut of his/her jib"

a judgmental expression about someone's appearance, character, or demeanor.

> The contour or expression of his face. A sailor's phrase. The cut of a jib or foresail of a ship indicates her character hence a sailor says of a suspicious vessel, he "does not like the cut of her jib."
>
> *The Brewer Dictionary of Phrase and Fable*, 1870

jibe (gybe)

the opposite of a TACK; to turn while going downwind so the wind goes to the LEE of the sails, causing the BOOM(s) to cross to the opposite side; the warning "Gybe-ho!" is recommended (as is ducking when you hear it). The word "jibe" carries with it the implication that the event is accidental; the word for a controlled jibe is WEAR.

Jolly Roger, the

a flag showing a white skull and crossbones on a black ground; and supposedly flown by PIRATES.

junk rig

a sail plan that uses fully battened lug-sails on an unstressed rig, sometimes on a partially stressed rig. Of oriental origin, a junk-rigged vessel has certain marked advantages: it is self-tacking (indeed, self-gybing); can be reefed by easing off the halyard; can be unreefed by hauling on the halyard; can be patched without affecting performance; does not require replacement sails; and looks cute when entering a new ANCHORAGE. Probably the best rig for single-handing and has excellent down-wind performance, but is pretty pathetic when beating to windward.

jury (jury-rig)

a temporary contrivance to repair rigging in order to get to the nearest safe haven; also used to describe a vessel that is short-canvassed. Broken masts are often rerigged to support staysails or cut-down jibs. Lost the RUDDER? Remove the HEAD(s) door, secure it to a spinnaker or jib-pole, and then lash the whole arrangement to the PUSHPIT and you have a jury rudder. All you have to do now is devise a jury door for the head(s).

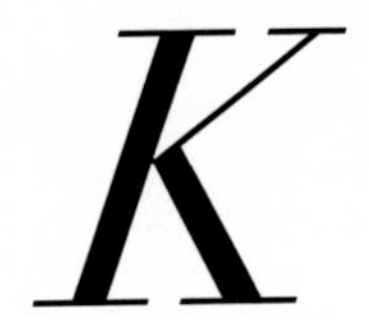

kannabis

see CANVAS.

Karachi Flyer

a little-known one-design racing dinghy unique to the Karachi Yacht Club, Pakistan; also the boat in which I learned to race a long time ago. A 14ft hard chine boat, it was reputedly designed during World War II by a Royal Air Force squadron leader and built locally. It 'planed with little urging and provided great sport in a very competitive class (there was little else to do but race three times a week). If there was a problem at all it was that the sun rotted the toe straps. The resulting backward somersault into the warm water of the bay always happened at a key moment in the desperate and close-fought battle for twelfth place.

kayak

the traditional decked CANOE of the Inuit people; made by stretching seal skins over a light frame. Modern versions constructed from GRP and ALUMINUM are used for sport and recreation.

kedge

a small secondary ANCHOR tied to the vessel via a rope warp (rather than warp-plus-chain) and usually stowed at or near the STERN. Although the kedge may be used as a second main anchor (and in some conditions you can never have too many anchors) its main purpose is in maneuvering the boat. Here is an example. You are heading into an unfamiliar haven when you run aground. Launch the dinghy and perch the kedge on the TRANSOM so you can kick it off with your foot. Then flake the warp on the sole and start rowing. When you are far enough out, drop the kedge and let the warp pay out. Back in the yacht, bend the warp around a winch, pray that the anchor is going to hold, and start winding. With any luck, it will pull you off. If not, put the kettle on and wait for the tide to flood again.

keel

for the builder, the keel is the main component of a boat's hull, providing the "backbone" from which the forming ribs project. For yachtsmen, the keel is the component of a sailing boat that projects from below the hull and carries the ballast used to counter the heeling moment caused by the effect of the wind on the sails—it keeps the boat upright in the water. Also serves to minimize the sideways movement of the boat. Keels are the subject of acute attention on the part of designers and, consequently, there are many variations, including the LONG KEEL, BILGE KEEL, FIN KEEL, FIN-AND-BULB KEEL, SWING KEELS, and CENTER BOARDS (center plates).

"keel, get back onto an even"

to straighten a boat when it has gone aground or is HEELING too much as a result of being over-canvased; now a vernacular expression meaning to put matters right, to regularize something.

keel-hauling

Another application of the ever-versatile YARDARM.

> Metaphorically, a long, troublesome, and vexatious examination or repetition of annoyances from one in authority. The term comes from a practice that was formerly common in the Dutch and many other navies of tying delinquents to a yardarm with weights on their feet, and dragging them by a rope under the keel of the ship, in at one side and out at the other. The result was often fatal.
>
> *The Brewer Dictionary of Phrase and Fable*, 1870

ketch

a yacht with two masts; the smaller one to the STERN of the MAINMAST is called the MIZZEN. On a ketch, the mizzen is set forward of the RUDDER. If the mizzen is stepped aft of the rudder, then the yacht is a YAWL.

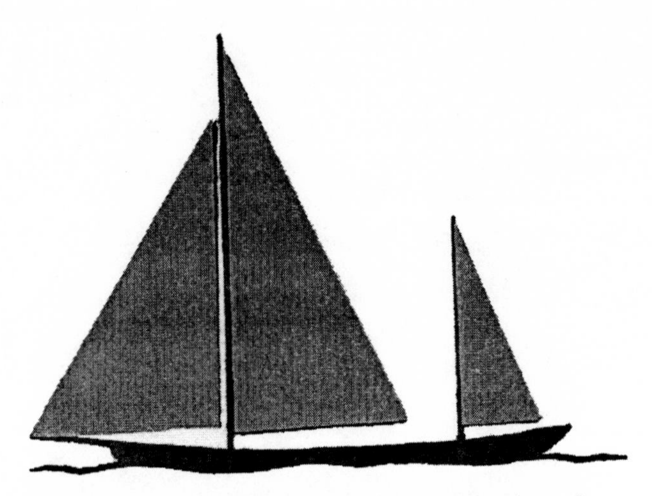

It is also related to the SCHOONER, as Jack London explained.

> The rig of the *Snark* is to be what is called the "ketch." The ketch rig is a compromise between the yawl and the schooner. Of late years the yawl rig has proved the best for cruising. The ketch retains the cruising virtues of the yawl, and in addition manages to embrace a few of the sailing virtues of the schooner. The foregoing must be taken with a pinch of salt. It is all theory in my head. I've never sailed a ketch, nor even seen one. The theory commends itself to me. Wait till I get out on the ocean, then I'll be able to tell more about the cruising and sailing qualities of the ketch.
>
> *The Cruise of the Snark* by Jack London

Kevlar

a very versatile, manmade substance combining lightness with great strength; can be used to make masts or sails and bulletproof vests.

key (cay)

"key" (as in Key Largo) is a phonetic misspelling of "cay"; an islet or small island (or coral or rock) in the Caribbean; often uncultivated and uninhabited. From the Spanish *cayo* or "rock."

kicking strap (US: vang)

a MARTINGALE used to flatten a sail (usually the MAINSAIL) by pulling the BOOM downward; usually a line incorporating a BLOCK and adjustable from the cockpit, but on high-tech BERMUDAN RIGS may be a metal spar. Called a "vang" in North American waters.

killing fish

When you get your first beautiful glistening tuna over the side you are going to have to kill it (unless you are Japanese, in which case you will want to eat it while it is still fresh).

Basically, you need to get firm hold of the fish on the DECK or cockpit sole and hit it a sharp blow over the head with something heavy; a winch handle will do. If you cannot stomach this, John Beatty has a solution:

> I had heard from someone in Tenerife of a kind way to kill fish and decided to give it a go. A fish uses its gills to extract oxygen from water in much the same way as a human lung extracts oxygen from air. I fetched a bottle of whisky and poured a small amount on to its bright red gills. The results were dramatic! The fish (a 20lb-plus tuna) quivered and died very peacefully in less than 10 seconds—no blood or guts. Although I can't be sure, I think that fish died as happy as any fish can.
>
> *The Breath of Angels* by John Beatty, 1995

I duly note that Mr. Beatty, who is Irish, did not use *Irish* whiskey for these last rites.

kill switch

a switch with a LANYARD that automatically shuts off an engine (usually an OUTBOARD) if pulled; also seen connected to wrists of people riding PERSONAL WATER CRAFT.

king spoke

the uppermost, vertical spoke of a ship's wheel; usually marked with tape or cord rather than a small crown.

knock-down

to be CAPSIZED by a sudden squall or a heavy breaking sea.

knot

one NAUTICAL MILE (1.15 miles or 1.85km) per *hour*. A knot is a measure of *speed*, not distance. Therefore "one knot per hour" makes no sense unless you are inventing a totally new and useless measure of acceleration.

knotmeter (UK: log)

see LOG.

knots

an excellent means of giving up smoking.

On the HARD you can sit watching TV, wishing you were on the water, while you constantly tie and retie knots. That way you can become so proficient you can tie the knot quickly and reliably without being able to see it. On deck at night, you can do the same thing but instead of watching TV, brush up on your ability to spot navigation stars and listen to early Miles Davies on your MP3 player. At the same time you are giving up smoking. There are plenty of excellent books available on how to tie knots, no excellent books on how to give up smoking. Incidentally, I have seen folks on yachts have postprandial knot-tying competitions; makes a change from Trivial Pursuit, I suppose (especially as I always get them with the running BOWLINE tied behind my back).

L

lagan

anything thrown overboard from a boat with a BUOY attached so it can be recovered later; old trick used by smugglers (and you thought those yellow buoys had lines of lobster pots on the end?).

lagoon

a bay formed by CORAL REEFS.

LANBY

Large Automatic Navigational Buoy. A super, all-purpose BUOY developed by Trinity House; employs light and sound signals with a radar beacon. With a height of between 12m and 15m, can be seen for about 15 NAUTICAL MILES.

landfall

the very first sight of land after a passage. It could be argued that only seafarers appreciate the real significance of this event. JOSEPH CONRAD was a seafarer and he captured the essence of it in the opening to *The Mirror of the Sea*.

> Landfall and Departure mark the rhythmical swing of a seaman's life and of a ship's career. From land to land is the most concise definition of a ship's earthly fate.

A "Departure" is not what a vain people of landsmen may think. The term "Landfall" is more easily understood; you fall in with the land, and it is a matter of a quick eye and of a clear atmosphere. The Departure is not the ship's going away from her port any more than the Landfall can be looked upon as the synonym of arrival. But there is this difference in the Departure: that the term does not imply so much a sea event as a definite act entailing a process—the precise observation of certain landmarks by means of the compass card.

Your Landfall, be it a peculiarly-shaped mountain, a rocky headland, or a stretch of sand-dunes, you meet at first with a single glance. Further recognition will follow in due course; but essentially a Landfall, good or bad, is made and done with at the first cry of "Land ho!" The Departure is distinctly a ceremony of navigation. A ship may have left her port some time before; she may have been at sea, in the fullest sense of the phrase, for days; but, for all that, as long as the coast she was about to leave remained in sight, a southern-going ship of yesterday had not in the sailor's sense begun the enterprise of a passage.

The taking of Departure, if not the last sight of the land, is, perhaps, the last professional recognition of the land on the part of a sailor. It is the technical, as distinguished from the sentimental, "good-bye." Henceforth he has done with the coast astern of his ship. It is a matter personal to the man. It is not the ship that takes her departure; the seaman takes his Departure by means of cross-bearings which fix the place of the first tiny pencil-cross on the white expanse of the track-chart, where the ship's position at noon shall be marked by just such another tiny pencil cross for every day of her passage. And there may be sixty, eighty, any number of these crosses on the ship's track from land to land. The greatest number in my experience was a hundred and thirty of such crosses from the pilot station at the Sand Heads in the Bay of Bengal to the Scilly's light. A bad passage. . . .

A Departure, the last professional sight of land, is always good, or at least good enough. For, even if the weather be thick, it does not matter much to a ship having all the open sea before her bows. A Landfall may be good or bad. You encompass the earth with one particular spot of it in your eye. In all the devious tracings the course of a sailing-ship leaves upon the white paper of a chart she is always aiming for that one little spot—maybe a small island in the ocean, a single headland upon the long coast of a continent, a lighthouse on a bluff, or simply the peaked form of a mountain like an ant-heap afloat upon the waters. But if you have sighted it on the expected bearing, then that Landfall is good. Fogs, snowstorms, gales thick with clouds and rain—those are the enemies of good Landfalls.

The Mirror of the Sea by Joseph Conrad, 1903

The common reaction is somewhat less philosophical; am I where I am supposed to be?

landmark

a prominent object or geographical feature used for navigation; may be natural (a headland), artificial (LEADING MARKS) or opportunistic (the church on the headland).

lanyard

a short cord attached to an item of equipment at one end and to a crew member or the boat at the other; used to prevent the object becoming separated from the aforementioned crew or boat.

lapstrake (US)

a method of boat construction involving the over-*lap*ping of long planks of wood (*strakes*); see also CLINKER-BUILT (UK). The method was originally developed when boats had to be built using timber that did not have a particularly straight edge and consequently would have been a nightmare to CAULK. *Or* used as a method of construction whereby the skin was built first and then the ribs added, if you want another view on the issue.

lash

to secure with a rope. "The dinghy was lashed to the deck" does not mean that you were upset with your dinghy.

lateen rig

a large triangular sail, the luff of which is laced to a long yard rigged at an angle to the mast, the lower end forming the TACK. A lateen sail flies LOOSE-FOOTED.

From the French *voile latine* (latin sail) supposedly because of its extensive use in the Mediterranean—though I have only ever seen them in the Red Sea and Indian Ocean.

lateral buoys

a system of BUOYAGE used to mark the left and right extents of a safe passage: PORT lateral BUOYS are typically cylindrically (can) shaped, are RED in color and may show a red light; STARBOARD lateral buoys are typically cone-shaped, are GREEN or black in color and may show a green light. See IALA and DIRECTION OF BUOYAGE because the Americas and Japan have a different arrangement from the rest of the world.

latitude

notional parallel lines of equal angle north and south of the EQUATOR; latitude 0° is the equator itself, 90° N is the North Pole, 90° S is the South Pole. In modern navigation, latitude is expressed in degrees, minutes and tenths of a minute (not seconds). When combined with a line of LONGITUDE, any position on the surface of the earth can be described.

The word itself comes from the Latin *latus*, broad. This it seems is because old charts were "portrait" format rather than the "landscape" format used today. So, the broader part of the chart (north–south) was measured in latitude.

lay

1. the pattern or "twist" of a three-stranded rope; see also MARRY.
2. a course can be "laid" when it can be held without the need to TACK.
3. (US) a particularly enjoyable human activity; done best in a BUNK (in a HAMMOCK at your own risk). See MARRY.

laying up

1. the process of making a rope by twisting the strands together.
2. the process of getting the boat out of the water for the winter (and the fresh coat of ANTI-FOULING).

lazarette (lazaretto, lazaret)

a cabin or other space used for stowage (usually toward or in the STERN of a boat, sometimes under the COCKPIT); this concept seems to have lapsed from the vocabulary of yacht designers, irritating inaccessibility beneath bunks seeming to be the main criteria for allocating such space these days.

leading edge

another name for the LUFF of a sail; or in popular parlance to be in the vanguard of some field or other: "She was at the leading edge of the advertising industry."

leading light

see LEADING MARK; now used in everyday English to indicate that somebody is prominent is a particular organization or field of activity: "He was the leading light of the bowling club."

leading mark

a MARK (or more usually a pair of marks) on shore used to indicate the safe passage into a harbor or other enclosed water; when two marks are used (the distant one higher than the closer one), they need to be in TRANSIT (in line) as the boat enters or leaves. Leading marks may also incorporate lights for use at night.

lead line (hand line)

a line used to determine the depth of water a boat is passing over (usually just before it goes aground); traditionally a lead weight of 7lbs (a "sounding lead," about 3kg) connected to a line marked off with ties every FATHOM.

> The proper way to mark a hand line is, black leather at 2 and 3 fathoms; white rag at 5; red rag at 7; white strip of leather, with a hole in it, at 10; and 13, 15 and 17 marked like 3, 5 and 7; two knots at 20; 3 at 30; and 4 at 40 with single pieces of cord at 25 and 35.
>
> R. H. Dana, 1855

The preferred technique is to swing the lead ahead from the BOW and to let it settle before hauling it back again; the preferred technique for a shirker was to just "swing the lead" and make a guess, thus saving the effort of pulling the lead back. "Plumbing the depths" also comes from this source. The pseudonym of American author Samuel Clemens is said to come from the riverboat cry "By the mark, twain!" (two fathoms) made to the helmsman. On leaving the Covent Garden Opera House in London after experiencing his first opera, Clemens is reputed to have remarked, "Apart from all the singing it was wonderful."

league

a measure of distance or, more precisely, *measures* of distance. I seem to remember being told at school that one league equaled seven statute miles.

However, in the Age of Exploration it seemed that the length of a league was pretty much optional; it could be whatever you wanted it to be.

The Portuguese Maritime League (PML) of 3.2 nautical miles (nm) was in common use at the time among Spanish as well as Portuguese seafarers. It was assumed that Columbus, being under contract to the Spanish crown, would have used this measure. However, a few years ago, an academic checked the log of some of his measured passages along the north coast of Cuba and deduced that Columbus's league must have been 1.5 nm in length. Sadly, though, there is no historical reference to such a measure. Coincidentally, this and other research did conclude that he was *not* using the PML. Later calculations based on the logs conclude that Columbus must have been using the Italian League, sometimes known as the Geometric League of 2.67 nm.

Given that little bit of history I supposed we should be quite relieved to be working with only the nautical mile, the statute mile, and the kilometer in the 21st century. Once Britain and the US get their acts together we will be able to drop the obsolete statute mile.

learn the ropes

on sailing vessels, rookies (or newbies) had to "learn the ropes"; but only rookies were likely to use the word "rope" when they were learning the function and purpose of each *line*.

lee

away from the direction of the wind, in shelter; from the Dutch word *lij* for "shelter."

lee boards

1. boards fitted to the sides of flat-bottomed sailing boats to reduce LEEWAY; usually tear-shaped and hinged so their depth can be varied; lee boards do the same job as the KEEL on a yacht but cannot provide righting BALLAST.
2. boards or sail cloth fitted to bunks to stop the occupant rolling out.

lee helm

the annoying tendency of some sailing boats to turn their BOW away from the wind; a propensity that has to be corrected by holding the helm toward the lee. Opposite of WEATHER HELM.

lee shore

a shore *toward* which the wind is blowing; in other words, a shore to the lee of the boat. It is advisable to have plenty of OFFING from one of these in a STORM. The terror of being in such a situation is recounted by Jack London.

> [A]bout the liveliest eight days of my life were spent in a small boat on the west coast of Korea. Never mind why I was thus voyaging up the Yellow Sea during the month of February in below-zero weather.
>
> The point is that I was in an open boat, a sampan, on a rocky coast where there were no light-houses and where the tides ran from thirty to sixty feet.
>
> My crew were Japanese fishermen. We did not speak each other's language. Yet there was nothing monotonous about that trip. Never shall I forget one particular cold bitter dawn, when, in the thick of driving snow, we took in sail and dropped our small anchor. The wind was howling out of the northwest, and we were on a lee shore. Ahead and astern, all escape was cut off by rocky headlands, against whose bases burst the unbroken seas. To windward a short distance, seen only between the snow-squalls, was a low rocky reef. It was this that inadequately protected us from the whole Yellow Sea that thundered in upon us.
>
> The Japanese crawled under a communal rice mat and went to sleep. I joined them, and for several hours we dozed fitfully. Then a sea deluged us out with icy water, and we found several inches of snow on top the mat. The reef to windward was disappearing under the rising tide, and moment by moment the seas broke more strongly over the rocks. The fishermen studied the shore anxiously.
>
> So did I, and with a sailor's eye, though I could see little chance for a swimmer to gain that surf-hammered line of rocks. I made signs toward the headlands on either flank. The Japanese shook their heads. I indicated that dreadful lee shore. Still they shook their heads and did nothing. My conclusion was that they were paralysed by the hopelessness of the situation. Yet our extremity increased with every minute, for the rising tide was robbing us of the reef that served as buffer. It soon became a case of swamping at our anchor. Seas were splashing on board in growing volume, and we baled constantly. And still my fishermen crew eyed the surf-battered shore and did nothing.
>
> At last, after many narrow escapes from complete swamping, the fishermen got into action. All hands tailed on to the anchor and hove it up. For'ard, as the boat's head paid off, we set a patch of sail about the size of a flour-sack.

> And we headed straight for shore. I unlaced my shoes, unbuttoned my great-coat and coat, and was ready to make a quick partial strip a minute or so before we struck. But we didn't strike, and, as we rushed in, I saw the beauty of the situation. Before us opened a narrow channel, frilled at its mouth with breaking seas. Yet, long before, when I had scanned the shore closely, there had been no such channel. *I had forgotten the thirty-foot tide.* And it was for this tide that the Japanese had so precariously waited. We ran the frill of breakers, curved into a tiny sheltered bay where the water was scarcely flawed by the gale, and landed on a beach where the salt sea of the last tide lay frozen in long curving lines. And this was one gale of three in the course of those eight days in the sampan.
>
> "Small-Boat Sailing" from *The Human Drift*
> by Jack London, 1911

That is one of those "And with one bound he was free!" stories. See SAMPAN.

leech

the aftmost edge of any sail.

leeward (pron. "loo-w'd")

the side of a sailing boat presently away from the wind; see WINDWARD. If you are feeling SEASICK, make sure you know which side this is.

leeway

the lateral distance traveled off-course as a result of the sideways pressure of the wind. Leeway varies from yacht to yacht, but skippers are usually aware of the allowance that needs to be made when calculating a course to steer (typically 5–10°). Minimized by the use of CENTERBOARDS, LEE BOARDS, and KEELS. "Having enough leeway" is now a vernacular expression meaning "having freedom" or "having enough room to make a mistake without dire consequences."

"left high and dry"

AGROUND; stranded.

length over all (LOA)

the overall length of a vessel; see MEASUREMENTS.

let fly

to release a SHEET, allowing a sail to fly free; in everyday English can be used to mean "lose one's temper," "to let fly at somebody."

lie to

a sailing boat which has stopped, almost head-to-wind, is said to be "lying to"; another touch of the tiller and she will be IN IRONS.

lifeboat

1. a boat used by coastal rescue services to deliver assistance to vessels and persons in distress at sea; especially in the UK where the service is an admirable volunteer organization called the Royal National Lifeboat Institute (RNLI).
2. a boat designed for the emergency evacuation of passengers and crew from ships; originally large, sturdy wooden rowing boats, more recently covered fiberglass or aluminum hulls fitted with diesel engines and commonly able to carry as many as eighty casualties, sometimes more. Lifeboats are launched by means of DAVITS.

In the early days of recreational sailing the latter type of lifeboat often provided the starting point for a handy sailing DAY-BOAT. The English lawyer and writer E. F. Knight took his from Hammersmith (up-river from central London), across the North Sea to Copenhagen in the Baltic. In Chapter 1 of his very readable account of the voyage, he describes the *Falcon* and her genealogy.

> The yacht at Hammersmith possessed two qualities not usually found together. She was of very light draught and yet she was an excellent sea-boat. She drew something under three feet, and so could enter the shallowest Danish boat-harbour. With her if I saw a port before me I could run in boldly, not needing a pilot, and without troubling my head about the depth of water; for, where any other boat had gone before mine was able to follow. She also looked like a craft that would put up with a good deal of heavy weather, and could be trusted to carry one safely across the North Sea. I saw that she was, in short, the very vessel I required; so I came to terms with her owner, and soon found that I had no reason to be disappointed with my bargain.
>
> The *Falcon*—for so I named her after my former vessel—was an old P. and O. life-boat, and had doubtless made many a voyage to India and back on a steamer's deck. As is the way with life-boats, her bow and stern were alike, and she had far more sheer than is ever given to a yacht. She had been built in the strongest manner by

> the well-known life-boat builder White, of Cowes. She was double-skinned, both skins being of the best teak, the outer of horizontal, the inner of diagonal, planking.
>
> The gentleman from whom I bought her had converted her into a yawl, or, to be more correct, a ketch, for her mizen-mast [*sic*] was well in-board, so that her mainsail was smaller and her mizen larger than is the case with yawls (an advantage as far as handiness is concerned). The water-tight compartments had been taken out of her, a false keel had been fastened on, and she had been decked all over with the exception of a small well. There was no appliance for covering over this well in bad weather, but I have never seen a pint of water tumble into it, so buoyant and admirable a sea-boat did the little vessel prove to be.
>
> The *Falcon* is jury-rigged; too much so indeed, her spars and sails being rather too small. Her mainmast lowers on a tabernacle, a system which I do not like for sea-work, but which proved useful on the Norfolk Broads. She is twenty-nine feet long and of three tons register.
>
> *The Falcon on the Baltic* by E. F. Knight, 1892

Sadly, however, Chapter 2 of the book is somewhat chillingly titled "The New Boat Leaks."

life jacket (US: personal flotation device or PFD)

a "garment" designed to keep a person in the water afloat until they can be rescued. A life jacket (sometimes "life vest") is not to be confused with the kind of "buoyancy aid" worn by racing dinghy sailors—a real life jacket should keep an unconscious victim afloat, with their head out of the water. Given that most people who die after falling overboard die from exposure rather than drowning, it could be argued that the only useful function for a life jacket (PFD) or BUOYANCY AID is to help the rescue services find your body. If the thing does not keep you afloat, it is not even performing that function.

Life jackets were originally solid arrangements ("Mae Wests") using cork packed into pouches extending up the chest, around the neck, and down again. These were far too bulky for moving around a yacht (or flying an aircraft) and were soon made inflatable, firstly by lung power and then by gas bottle for smokers. Modern life jackets will inflate automatically when coming into contact with water. They come with a flashing light (the strobe type is best) so you can be seen in the dark and a whistle for attracting attention. The main thing to consider when choosing a life jacket is (in Europe) to ensure that it conforms to the CE standard. This is based on a minimum lift force of 150 newtons. But 275-newton jackets cost only a little more. A recommended optional extra is a spray hood. Many "overboard" victims drown from the spray thrown up in heavy weather (try this in a swimming pool by having someone splash water into your face—you will soon find it hard to catch your breath). The hood is made of clear plastic and folds out from the back of the jacket right down over your face. Also check any tests carried out by sailing magazines and especially look to see if they made sure that the jacket would turn you onto your back while unconscious. Life jackets can also be incorporated into a SAFETY HARNESS. See also SURVIVAL SUITS.

It is important for skippers to take a firm lead on the wearing of life jackets. Being cavalier about your personal safety has nothing to do with bravery and everything to do with congenital stupidity. From Force 6 and up, a jacket and/ or (clipped-on) safety harness should be worn when outside the cockpit. From Gale Force 8 and above, both life jacket and harness should be worn—even in the cockpit. This fairly minimal rule should apply to everyone, regardless of their experience.

On 14 June 1998, news came through on my radio of the sad death of the famous French yachtsman Eric Tabarly. It seems that he went overboard while helping reef the MAINSAIL on *Pen Duick* in the Bristol Channel off Haverfordwest, Wales. Weather conditions were not over-severe. Members of his crew reported that he was not wearing a life jacket at the time.

lifeline

a strong wire that extends around the outside of a yacht's DECK. Held up by metal STANCHIONS, the lifelines (there may be one or two) effectively make a rail to prevent crew from falling over the side. Netting of light cord is sometimes fitted between the lifelines to prevent sails, buckets, children, and crew members going overboard. Essential for hanging out the laundry.

life-tender

The typical inflatable LIFERAFT is quite small when deflated and packed into its canister or valise. And that is where the plaudits run out. Being inside one on the water is not an experience I would want to repeat. They are uncomfortable (especially before being bailed out and dried out), claustrophobic, and unnavigable—they go nowhere fast.

This is an important psychological point. If you are in coastal waters and are properly equipped with EPIRB, flares, rockets, smoke, and dye, you are going to get picked up fairly soon, *Insh' Allah*. But the further offshore you are, the less likely is this dream scenario going to become reality. So, what do you do? Just lie there and pray? Or are you going to try and get yourself out of the predicament?

To do the latter you need a different type of life craft, a navigable escape module. But, before we go any further, let us consider the inflatable tender. Unless you have a permanent berth in a marina and never go anywhere else, then you need a tender. It needs to be inflatable or, if solid in construction, small enough to fit on the boat and in the boot (trunk) of the car or on its roof. Whether rowed or powered by a small outboard, it will provide the essential ferry service between the yacht and the shore for provisions and crew.

Far more people fall overboard from tenders in "havens" than from sailboats on the high seas. They are, after all, easy craft to fall out of, especially when badly trimmed or in severe weather conditions. The particular problem with solid-hulled craft is that they often have no in-built buoyancy; capsize one and it will sink. You could, of course, add some buoyancy to it and thus reduce its tendency to capsize and eliminate its tendency to turn into a submarine. Sounds like a liferaft, doesn't it?

The other piece of overwhelming logic influencing this design process is the fact that, when in harbor you don't need a liferaft and, when at sea, you don't need a tender. So why have two craft?

A British company called Henshaw Inflatables has come up with a solution. They provide an optional conversion kit for their TINKER dinghy. This enables you to deflate the dinghy when ready to set sail. The dinghy is packed up with an inflatable canopy, a survival pack, and a gas auto-inflate system. Throw this over the side and it will inflate and provide you with a liferaft that can be rowed, sailed (even), and outboard-powered.

The only problem I have with this fine solution is that I really would prefer a solid-hulled "harbor-taxi." And a solid-hulled "escape module." First preference would be GRP, then wood. Obviously GRP is easier to maintain but both are fairly straightforward to modify.

This is the kind of project I would undertake on a good, second-hand GRP tender (I have done it on a new one). The money you save at this stage will pay for the modifications. Do not buy one that is too small. The test is this: if the tender does not have a center THWART (seat), could you lie down in it? Even allowing for storage space in the BOW and, maybe, under the stern thwart? Here are some of the modifications you could make (after ensuring the integrity of the hull, of course).

- Make sure that the tender is unsinkable. Do this by adding buoyancy along both sides from the TRANSOM to the BOW and across the STERN. Do not use the space within the hull because you are going to need all of this. The buoyancy can be inflatable or pre-inflated. Sausage-shaped rollers for launching and recovering boats are usually about the right length and diameter. Secure them to the boat inside tubes of UV-resistant cloth. When needed—either in an emergency or in rough anchorage conditions—they can be inflated using a lot of puff (slow), with a 12v tire inflator (not quite so slow), or with the type of CO_2 gas cylinder used to inflate liferafts (the fast-but-unrepeatable option). The pre-inflated approach is to use cylindrical FENDERS. As you leave a berth one of the first things you do is recover the fenders from the rails. Instead of stowing them in the lazarette, stuff them into the canvas tubes of the life tender. Neat, eh?
- A different option might be considered for the stern. Small boats are as difficult to get into as a celebrity nightclub. In view of this, I think the stern buoyancy should be in the form of a permanent step fitted to the lower half of the TRANSOM. Then securely fit a rope ladder (between 1m and 2m in length) and stow it using bungee cord on the step.
- If you have space in the BOW, convert to a watertight locker. Do the same with the stern thwart. Take out the present center thwart and convert it to a locker that can be removed and placed fore-and-aft when you need to lie down. Now you have permanent space for everything that used to go in your GRAB BAG. You can also ensure that your survival kit is far better than the one normally hidden in a LIFERAFT.
- Now for the means of propulsion. You are either going to row or to sail. Either way, you will need two stout OARS. When rowing, put the oars in the ROWLOCKS (securing each with a short line) and start rowing. If you are going to sail, you will need a mast. One of the oars will do for this. Drilling holes into the end of the blade (do this with both oars) will enable you to add fittings for SHROUDS and a HALYARD. The other end of the oar will need to fit into the KEEL. As far as rigging is concerned, you can opt for either a SQUARE RIG, or a LUG RIG of some kind. Either way, you will need a spar and maybe even an improvised LEEBOARD or two.
- Steering can be achieved by using the second oar in a SCULLING hole at the top of the TRANSOM.
- Consider taking a third oar; that way, if you lose one, you will still be in business. Trouble finding a third oar? Not really—just find someone who just lost one.
- Finally, you need shelter. This is essential for keeping off both sun and rain. Decide in advance how you are going to rig this and make sure the canvas

is cut to shape and has rivet-lined holes in the right places. (It may be that the canopy can be used for rainwater catchment, but I haven't quite figured that out yet.)

When everything is looking good (but before it is finished) take your new life-tender out for sea trials. Then the only remaining problem is finding somewhere to stow it. I would suggest DAVITS off the STERN (if you have a flattish transom) or inverted on the cabin roof or FOREDECK. In the latter cases, have a permanent RIG that everyone can use for launching and recovery.

Anyone want to go into business making these things?

light characteristics of buoys

the characteristics of lights on BUOYS are shown on modern charts using the following shorthand (note that the range of a light is also an important aid to recognition).

Fixed	F	A constant light
Flashing	Fl	Single-flashing
	LFl	Long-flashing
	Fl(2)	Group-flashing
	Fl(2+1)	Composite group-flashing
Quick	Q	Continuous quick
	Q(4)	Group-quick
	IQ	Interrupted quick
Very Quick	VQ	Continuous very quick
	VQ(3)	Group very quick
	IVQ	Interrupted very quick
Ultra Quick	UQ	Continuous ultra quick
	IUQ	Interrupted ultra quick
Fixed and Flashing	FFl	A fixed light that flashes
Occulting	Oc	Single-occulting
	Oc(3)	Group-occulting
	Oc(1+2)	Composite group-occulting
Isophase	Iso	Light and dark equal
Alternating	Alternating colors (e.g., Al.WR)	Alternating white/red
Morse Code	Mo(K)	Morse letter "K"

lighter sockets

there are a lot of GADGETS that are made for cars (cell phone chargers, fans, lights, vacuum cleaners) that can be used in boats. They have the advantage of cheapness, being made for a much bigger market, and not being sold through chandlers. Such devices are powered from 12v supplies through the lighter socket (cigar lighters, as they are sometimes called). Having such a socket on a boat does, therefore, make the use of these gubbins possible for seafarers. When buying, choose one that has a shield and a cover designed to keep the water out. This is less important if the socket is installed in the OFFICE, but essential if it is anywhere on DECK.

lights on sailing vessels

the lights used on a sailing boat are important and are governed by maritime law. International Rule 25 of the COLREGS (US: NAVRULES) says this:

> a) A sailing vessel underway shall exhibit:
> (i) sidelights;
> (ii) a sternlight.
>
> b) In a sailing vessel of less than 20 meters in length the lights prescribed in paragraph (a) of this Rule may be combined in one lantern carried at or near the top of the mast where it can best be seen.
>
> c) A sailing vessel underway may, in addition to the lights prescribed in paragraph (a) of this Rule, exhibit at or near the top of the mast, where they can best be seen, two all-round lights in a vertical line, the upper being red and the lower green, but these lights shall not be exhibited in conjunction with the combined lantern permitted by paragraph (b) of this Rule.
>
> d) A sailing vessel of less than 7 meters in length shall, if practicable, exhibit the lights prescribed in paragraph (a) or (b) of this Rule, but if she does not, she shall have ready at hand an electric torch or lighted lantern showing a white light which shall be exhibited in sufficient time to prevent collision.
> (i) A vessel under oars may exhibit the lights prescribed in this Rule for sailing vessels, but if she does not, she shall have ready at hand an electric torch or lighted lantern showing a white light which shall be exhibited in sufficient time to prevent collision.
> (ii) A vessel proceeding under sail when also being propelled by machinery shall exhibit forward where it can best be seen a conical shape, apex downwards.

The above rule includes names for lights which need some clarification. This is provided by International Rule 21 as follows:

Definitions

a) "Masthead light" means a white light placed over the fore and aft centerline of the vessel showing an unbroken light over an arc of the horizon of 225 degrees and so fixed as to show the light from right ahead to 22.5 degrees abaft the beam on either side of the vessel.
b) "Sidelights" means a green light on the starboard side and a red light on the port side each showing an unbroken light over an arc of the horizon of 112.5 degrees and so fixed as to show the light from right ahead to 22.5 degrees abaft the beam on its respective side. In a vessel of less than 20 meters in length the sidelights may be combined in one lantern carried on the fore and aft centerline of the vessel.
c) "Sternlight" means a white light placed as nearly as practicable at the stern showing an unbroken light over an arc of the horizon of 135 degrees and so fixed as to show the light 67.5 degrees from right aft on each side of the vessel.
d) "Towing light" means a yellow light having the same characteristics as the "sternlight" defined in paragraph (c) of this Rule.
e) "All-round light" means a light showing an unbroken light over an arc of the horizon of 360 degrees.
f) "Flashing light" means a light flashing at regular intervals at a frequency of 120 flashes or more per minute.

lignum vitae

a hard, resinous wood often used in the form of friction blocks in the complicated running rigging of a JUNK-RIGGED vessel; from the Latin for "wood of life." Lignum vitae supposedly has medicinal properties, but I have only ever discovered the therapeutic value of sitting on the boat in the sun, rubbing sump oil into it.

limber holes

holes beneath the CABIN SOLEs in the cross-members and BULKHEADS that enable bilge water to slosh toward that part of the hull in which the pump is located, the PUMP WELL or SUMP.

line

a rope, cord or cable on a ship; as in "throw me a line," "the shorelines were tossed onto the deck."

Line, the

The equator.

list

a vessel is said to list when it leans to one side because the distribution of weight within its hull is unbalanced (e.g., shifting cargo); can sometimes occur on a sailboat if the fresh water and fuel tanks are placed in opposite BILGES, but do not empty evenly.

live-aboard

a "live-aboard" is someone who resides on a boat; see OCEAN GYPSIES.

livestock

such was the difficulty of provisioning with fresh food on long voyages that animals would be carried onboard and slaughtered as required en route. This floating menagerie somewhat alarmed Miss Emily Brittle during her long voyage to India.

> The bulkheads of cabins were constantly creaking,
> In concert with pigs, who as often were squeaking;
> Such a clatter above from the chick to the goose,
> I thought the livestock on the poop had broke loose;
> Dog, puppies and monkeys of ev'ry degree,
> Howled peals of loud discord in harsh symphony,
> Whilst near to my cabin a sad noisy brute
> Most cruelly tortured a poor German flute.
>
> from *The India Guide; or, Journal of a Voyage to the East Indies in the Year MDCCLXXX*
> by Sir George Dallas (1758–1833)

See also SEA LEGS and FIDDLE.

Lizard

1. a rope with a THIMBLE spliced into one end so it can run along another rope.
2. the Lizard is a promontory on the south coast of Cornwall, England; see the MANACLES.

LOA

see LENGTH OVER ALL; MEASUREMENTS.

local knowledge

quayside wisdom; always listen to the local knowledge.

> It was on a stifling hot beach in the Caribbean that I tore up my resolution to retire from dinghy sailing. The tree-covered island in the bay seemed to have caves at sea level, seemed to have mystery; exploration was called for. The hotel had a selection of dinghies and chose one a bit bigger than a Topper and smaller than a Fireball. It was a bit like a plastic ironing board, but had a jib. The attendant rigged it as I watched.
>
> "You know how to sail one of these, man?"
>
> "Sure," I said, sipping my rum punch and adjusting my Ray-Bans®.
>
> "It'll put your back out again," said my companion, smugly, as she rubbed more suntan oil onto her belly. We slid the boat into the water; I stuck a foot in and picked up the mainsheet. The attendant gave an unexpected shove as I pushed against the sand and I ended in a heap in the cockpit (such as it was). The wind was onshore, so the dinghy went backward. The attendant shoved again and this time I managed to sheet in the main and cleat it. The dinghy heeled over.
>
> "Don't go round the island, man!" the attendant called.
>
> "OK," I replied as I got my feet under the straps and countered the heel. I pulled in the jib sheet and leaned out some more. We were under way, but . . . I eased up into the wind so I could get the centerboard and rudder fully down. Now we were in business. I drove her hard, leaned out flat, my aging stomach muscles taking the strain. I looked back toward the beach, but no one seemed to be watching. Oh, well. The bow lifted to plane and the dinghy hissed along toward the island. I eased off and enjoyed the ride over the bright blue water. By cocking a leg over the tiller I freed my hands and lit a cigarette. Maybe I should have brought the rum punch? I made my way around the island. As I turned onto a reach I noticed a small villa nestling in a copse of purple bougainvilleas. I could smell their sweet blossoms as I turned again to head back between the island and the beach. I kept as close to the shore as I dared, but was well in the lee of the island.
>
> Something made me look up, I don't know what. There, ahead of me, was an electric power cable running out from the shore to just below the villa. I pulled hard on the jib and mainsheets to get some heel—but not enough. The mast was going to hit the cable! I dived across the dinghy and hung onto a shroud, my elbow skimming the water. The mast tilted and ducked under the humming wire. I dived back across the hull to stop the capsize and managed

to get back on course. What did the beach attendant say? Don't go round the island? When I got back to the hotel, he complimented my boat handling and ignored my stupidity. Ever grateful, I bought him a beer. As my companion and I headed back to our room for an afternoon nap, I felt a slight twinge in my lower back.

locals

On stopovers in developing countries, especially Africa, India, and parts of Southeast Asia, it can be a good idea have some items nonessential to the boat that can be used as free gifts to local people. Here are some suggestions:

- Fishing hooks; small, stainless steel ones about the right size for catching fish in coastal waters. Fishing line is an alternative, but this can be fiddlier to share out. Steel traces are also useful, but when you have bought a reasonable batch of these, it will start to appear in large letters in the budget.
- Domestic sewing materials such as needle and cotton.
- Polaroid photos seem popular with visitors who have been watching too many Hollywood movies. (And I sometimes think Westerners are more naively impressed by the technology than the locals are.) The usefulness of such things is also questionable. The same applies to pictures of your boat—really exciting. . . .
- Pencils and school notebooks. Deciding what will be useful requires a bit of research. It is amazing how many kids have to go to school in Africa with no writing implements. Go and chat with the teacher at the nearest school and ask what would be helpful. He or she will almost certainly say "pencils." If you are stocked up with these (and exercise books) or can buy them locally give them to the teacher to distribute.
- T-shirts are often appreciated and can be bought in bulk for as little as $1 each—but the bulk is the problem; they can take up far too much space on smaller boats. Medium size is best if you are going for this option—people in developing countries are rarely as fat as I—and small children's sizes can only be worn by the intended recipients, not their older siblings.
- And, talking about kids, small gift-packs of Legos have both durability and longevity of interest. I still have mine.
- Sample-sized soaps, shampoos, and other toiletries often go down well with the local girls, but have a female member of the crew hand these out; males may end up being chased down the beach by an angry boyfriend/husband with a machete.
- I have heard of an Australian yachtsman who sails around with a huge box full of discarded reading glasses. Brilliant.

- A pennant or similar with the yacht's name on is also an effective gift for the local bar or yacht club, if there is one, and if it has extended you a kindness such a free membership for your stay.

Obviously try to take account of local cultures and traditions. If in doubt, ask an expatriate for advice. Also watch the timing. Handing out goodies (especially the beads and mirrors) on arrival smacks of patronizing missionary zeal. Handing them out on departure might mean you miss out on the odd freshly caught fish or basket of fruit because they had got the impression you were tight-fisted.

local time

the time at the current positional of the boat; local time is always relative to GREENWICH MEAN TIME, 15° of longitude equaling one hour and subtracted when west of the GREENWICH MERIDIAN, added when east. Local time can varied by one hour of DAYLIGHT SAVING TIME.

locker

a storage compartment, a place to stow things; may have a lock, but not necessarily.

lodestone

a piece of magnetite (iron oxide). As its name suggests, this mineral is magnetic and, when hung freely from a peace of light cord or string, will tend to orientate itself north–south. An early, crude form of compass and useless for getting bearings.

log (US: knotmeter)

any device employed to measure the distance traveled through the water; an essential navigation aid.

A very common form of log, still in use, is one that involves trailing a small propeller behind the vessel. The propeller turns the line, causing the inboard end of the line to turn a series of gears that cause a numerical display of numbers to rotate (somewhat like the pre-digital distance meter on a car). The downside of this device is that the line has a tendency to get into an awful tangle. A later variation has a more sophisticated propeller that sends pulses along a wire connecting it to the "head" (usually clamped to the PUSHPIT), which counts the pulses and converts them into a digital display. Most yachts these days are fitted with log that works from a plastic propeller fitted through the hull where it grabs hold of every passing piece of seaweed. (Fortunately, you can clear or change these propellers from inside the hull.) In all three cases, two values are recorded and displayed; the distance traveled through the water since you first started using the log and the distance traveled on this trip. To bring the story up to date, you can now select DMG (Distance Made Good) on the GPS or other options such as the distance from the last WAYPOINT.

log-book

the official record kept of a vessel's voyage including date and times of departure and arrival, courses steered, weather conditions, and so forth.

loggerheads, loggerhead turtle

in the modern English vernacular, "to be at loggerheads" means to argue, to be in conflict. The "loggerhead" sea turtle (*Testudo carettais*) is seen around the world in warmer latitudes. Large specimens can weigh in at 450kg (over 1,000lbs). I used to live somewhere on the Indian Ocean where it was good to walk the beach in the moonlight and watch the turtles lay their eggs in deep holes in the sand, then to escort them back to sea before dawn. The carcasses of turtles who didn't make it littered the beach after being killed by the locals. "Loggerhead": a term of abuse for someone who is stupid. Such as an idiot who will kill a turtle just for the fun of it?

log-slate

this device predated the Post-it® sticker and other odd scraps of paper for communications between the watch officer and skipper or navigator. It enabled the off-going watch to chalk up the time of any changes in wind strength and direction as well as course changes adopted as a result. These could then be transcribed into the log proper for use in calculating a DEAD RECKONING. As E. F. Knight discovered en route from Southampton to South America in 1889 onboard his 64ft cutter *Alerte*, the use of a log-slate was not without its problems.

> While going down Channel we had kept watch and watch in the usual sea fashion, the first mate taking one watch and myself the other. But now that we were out at sea, clear of all danger, it became unnecessary to continue this somewhat wearisome four hours up and four hours down system; so we divided ourselves into three watches, the second mate taking the third watch. This gave the men an eight hours' rest below at a stretch, instead of only four. As we had three paid hands in addition to the cook, one of these was allotted to each

watch. But before reaching the South American coast the second mate resigned his post, and we reverted to the watch-and-watch system again, which was observed until the termination of the cruise.

A good deal of useless form was kept up at this early stage of the voyage. A log-slate was suspended in the saloon, and each officer as he came below would write up a full account of all that had occurred in his watch. The most uninteresting details were minutely chronicled only to be rubbed off the slate each midday—and I think there was a little disappointment expressed because I would not copy all these down in my log-book. Had I done so that log-book would have been a dreadful volume to peruse.

To us, however, the log-slate was a source of great amusement on account of its utter fallaciousness. The patent log was, of course, put overboard when we were making the land, but when we were out on the ocean and no land was near us we naturally did not take the trouble to do this, neither did we make use of the common log-ship or keep a strict dead reckoning. But, despite this, the officer of a watch would religiously jot down the exact number of knots and furlongs he professed to have sailed during each of his four hours on duty; he did not even try to guess the distance to the best of his ability; he was bred with an ambition to show the best record for his watch; so he would first scan the slate to see how many knots the officer just relieved boasted to have accomplished, and then he would unblushingly write down a slightly greater number of miles as the result of his own watch, quite regardless of any fall in the wind or other retarding cause.

Thus: if five knots an hour [*sic*] had been made in one watch, five and a quarter would probably be logged for the next, and five and a half for the next. Sometimes there was a flat calm throughout a watch, and then the ingenious officer, though he could not help himself and was compelled to write himself down a zero before three of the hours, would compensate for this by putting down a big number in front of that hour during which he imagined that all the individuals of his rival watches were fast asleep below, and would boldly assert in explanation that just then he had been favoured with a strong squall to help him along.

No one put any confidence in this mendacious slate, which soon became known on board as the "Competition Log," and inspired our wits with many merry quips. The distance made in each twenty-four hours as recorded by the Competition Log was about fifty per cent greater than that calculated from observations of the sun.

E. F. Knight, *Cruise of the "Alerte" in Search of Treasure*, 1890

Edward Frederick Knight was an English barrister who was not slow to mete out a little summary justice when dealing with his "paid crew": "He had done this on two previous occasions, also when under the influence of smuggled spirits, and had quickly been brought to his senses and to his work by having his head punched." (By the way, what do you call a lawyer in passage to South America? A reasonably good start.)

"long in the jaw"

originally used to describe a rope that has lost all its stretch; now in the vernacular to mean "past its/his/her sell-by date."

longitude

lines of longitude are great circles running through TRUE NORTH and representing angular distance from the GREENWICH MERIDIAN; in modern navigation, longitude is expressed in degrees, minutes, and tenths of a minute. When combined with a line of LATITUDE, longitude enables any position on the surface of the earth to be described in numerical terms.

The search for a practical means of determining longitude was one of the biggest scientific challenges of history; to the winning nation fell the prize of global exploration, enrichment, and imperial power. After centuries of fruitless effort, the English decided to throw money at the problem and passed into law the Longitude Act of 1714. This provided for

> a publick reward for such person or persons as shall discover the longitude . . . to a sum of ten thousand pounds, if it determines the said longitude to one degree of a great circle, or sixty geographical miles, to fifteen thousand pounds, if it determines the same to two thirds of the distance, and to twenty thousand pounds, if it determines the same to one half of the same distance.

In today's terms these sums represent lottery jackpots with a couple of rollovers. It is now common knowledge that clockmaker John Harrison won. In so doing, he finally put paid to a succession of remarkably stupid ideas.

Among these, the booby prize went to the 1687 proposal to use dogs as onboard timepieces. Yes, dogs. The idea went like this: cut a dog's leg so that it bleeds; wrap a bandage around the wound; when the next ship is ready to set sail to distant climes, put the dog onboard but keep the bandage. Do this for every ship leaving England. The soiled bandages would be entrusted to the care of a trusty timekeeper who would keep them close by an accurate pendulum clock. When an hour struck, the timekeeper would soak the bandage in a magic potion containing "powder of sympathy." As this was done, the wounded dog to which the bandage belonged would yelp loudly, warning the navigator to turn the ship's hourglass! The remarkable thing about this idea is that it betrays a thorough understanding of the relationship between time and longitude. Although it was a very effective solution to the stray dog problem it relied too heavily on the effectiveness of the mystical "powder of sympathy" to be a practical solution to the stray longitude problem. This however was an era when women were not allowed to attend scientific experiments in case their irrational brain patterns interfered with the results.

longshoreman

anyone who makes their living onshore but in support of the shipping industry.

loose cannon

a "loose cannon" was a very dangerous thing on the gun deck of a MAN O' WAR in a heavy sea.

loose-footed (footloose)

a description of any FORE-AND-AFT sail which is not fitted to a boom and is, therefore, loose at its FOOT and liable to FLOG in an uncontrollable manner; hence, "footloose and fancy-free."

Loran

an navigation system based on the use of radio beacons; see also DECCA and GLOBAL POSITIONING SYSTEM.

lubber line, lub line

a mark on a COMPASS which is in alignment with the CENTER-LINE of a vessel and, against the COMPASS CARD, gives the HEADING of the vessel.

luff

1. the "luff" is the leading edge of a sail; the edge fixed to the MAST in the case of the MAINSAIL and MIZZEN and the one fixed to the FORESTAY in the case of a JIB.
2. a boat is "luffed up" when she is turned to point closer to the wind.

luff rope, a

see BOLT ROPE.

lugsail (lugger)

a four-sided MAINSAIL, the FOOT of which is attached to a BOOM while the head is supported by a YARD (or lug) which is hauled up the mast by the HALYARD when the sail is set.

lure

a fancy arrangement concealing a hook and use to catch fish when TROLLING.

Few lures look anything like a baitfish in terms of shape or color but obviously look tasty enough to the stupid old tuna. If you don't get a result with one lure, try another—their effectiveness varies with sea conditions and the light. You can also vary the depth. Skipping the lure over the surface has been known to work, but you can try it a little deeper by adding a weight at the shipboard end of the TRACE.

lying to

see HOVE TO

M

mainsail (pron. "mains'l")

the primary sail of any boat. In a SQUARE-RIGGER this is the lowest sail on the main mast. On a BERMUDAN or lug-rigged vessel this is the boomed sail with its luff attached to the main mast. Mainsails used to provide the principal driving force of a yacht. On Bermudan rigs, this task has, on some points of sailing, been taken over by the GENOA or the SPINNAKER.

make fast to

to tie a line to something.

malaria

see MOSQUITO.

Manacles, the

a rocky reef off the east coast of the Lizard Peninsular in Cornwall, England. In the occasional desperate search for mackerel to use as shark bait, Robin Vinnicombe used to take *Huntress* right inside the Manacles. No big deal in the conditions at the time—other than the fact that I was onboard. It becomes quite alarming when you cannot see the way out. . . . (And what a wonderfully horrible name for a reef that has taken many seafarers over the centuries.)

manatee

a sea cow with a rounded tail flipper, living in shallow waters of the tropical Atlantic, threatened with extinction. See COCAINE.

man o' war

a man o' war was ship built for the purpose of combat at sea. It should be noted however that even trading vessels would carry some cannon for defensive purposes, especially against privateers and pirates. This is almost unheard-of in the 21st century; except in the case of tankers and container ships making passage through the pirate-infested waters of Southeast Asia and around the Horn of Africa. See PIRATES.

man overboard (MOB)

see PERSON IN WATER.

man rope

a rope to hold on to when going up or down a COMPANIONWAY or accommodation ladder; unclear what women are supposed to hold on to.

Marconi rig

a really tall mast used with a BERMUDAN SAIL PLAN; having a mast that supports lots of antenna is not good enough, I'm afraid. Named after Guglielmo Marconi, who certainly once held the world record for tallest (radio) mast stepped on the HARD.

Marie Celeste

The commonly used but wrong name for an American BRIG that was found mysteriously abandoned in the North Atlantic in 1872. Sir Arthur Conan Doyle was probably responsible for this; long before he started writing the Sherlock Holmes stories he published a weird speculation that the "*Marie*" *Celeste* had been hijacked by a "black power" terrorist group.

See MARY CELESTE.

Marigot Bay

On the west coast of St. Lucia in the Windward Islands, Marigot Harbor combines an exquisite natural beauty with being a pretty-near perfect hurricane hole. Surrounded by steep forested hills rising to 250m (c800ft), this tiny bay has become a popular ANCHORAGE for visiting yachtsmen (maybe too popular these days). Even if you have parked your vessel in the Rodney Bay MARINA to the north, Marigot Bay is still worth a visit for lunch or an evening drink. It is not too far by taxi or you could rent a 4WD and include it in a tour of the island. The bay is accessed via a narrow track off the main coast road that runs between the capital of Castries and the main airport at the southern tip of the island. (Do not take "Main Coast Road" too seriously—Pacific Coast Highway it ain't.) You park at the bottom of the track and take the free ferry across to the small hotel on the peninsula that projects into the bay from the north.

The first time I entered Marigot bay was from the sea. I was helming a beat-up old Baltic Trader. The way in is shallow, so you have to get it right. It is also so narrow that, from the stern of a boat as big as this, you lose sight of the channel altogether. So there I am, struggling to get it straight, trying to take in the breath-taking scenery at the same time, when I spot the Navigation Hazard. She was leaning against the mast of a 32ft French sloop (probably from Martinique), toying with what looked like a winch handle as they waited for us to pass through. Now it is not at all unusual for ocean passage makers to dispense with clothes in tropical latitudes, but in Marigot Bay?

A shout from the skipper in the BOW drew my attention to a 50ft gin palace ahead of us and the siren was suddenly gone from my life. It was only through an oversight, I am sure, that she was not listed in Notices to Mariners.

marina

an artificial haven for sailboats and motorboats about which yachtsman and novelist Sam Llewellyn once wrote

> I want to tell you from the bottom of my heart that in my view marinas are like car parks [parking lots], except that they contain more thieves and are less romantic.
>
> *Small Boat to Scotland*, Practical Boat Owner, 1997

Less romantic? What could he possibly mean?

marinate, to

literally, to "dunk in the sea." The original marinade was brine, salt water used for pickling food as a means of preserving it. Vinegar and soy sauce came along later.

marines

the most concise definition is probably "amphibious infantry"—although it would probably take a few more pages to describe the US Marine Corps, a part of the US Navy that is bigger than the whole of the British military. The USMC and the Royal Marines both have remarkable battle histories. The concept of putting fighting men on ships (they hate to be called "soldiers") may well predate the era of the Vikings. Just as warships need specialist sailors, they also need specialist fighters to help defend the ship at sea and to engage in onshore engagement with the enemy. Here is a shocking story about marines from the 18th century.

(Refers to the adventures of Commander John Byron. See also GIANTS OF PATAGONIA, MIDSHIPMAN, MUTINY, SHIP FEVER, and SQUADRON.)

In the late summer of 1740 the Honorable John Byron was posted to serve aboard HMS *Wager*. Originally an armed trading vessel, the *Wager* had been bought by the Royal Navy from the East India Company and fitted out for naval service as a FRIGATE at Deptford Dockyard on the Thames in southeast London. The *Wager* became part of a SQUADRON under the command of Commodore George Anson. The squadron comprised six warships, two transports carrying supplies, and 1,854 sailors and marines.

Led by the flagship, HMS *Centurion*, with 60 guns and 400 men, the squadron left its home port of Portsmouth on 18 September 1740. Although he must have been proud to lead a fleet under sail out into the English Channel, George Anson was experienced enough to know that trouble was looming.

The Admiralty had charged him with attacking the Spanish colonies on the Pacific coast of South America and that was going to require fighting men. The crews of all navy warships were trained to fire the vessels' cannons. But fit and well-trained marines were needed to board enemy ships and to go ashore in the boats and capture ports and garrisons. And that was what Anson didn't have. Britain was at war with Spain and fighting men were in short supply. Although the Admiralty had promised him 500 marines, their lordships began to cheat on the deal in the most loathsome way imaginable.

Chelsea Hospital in London was ordered to make available 500 *invalids* (pronounced the French way)—sick or wounded soldiers able to perform light duties. But the "Chelsea pensioners" got wind of what was in store for them and the ones who could sensibly made themselves scarce. Of the 259 unfortunate souls who were eventually delivered to the dockside many were on stretchers. That left a shortfall of 241. When the men making up the numbers arrived, they were standing upright and dressed like marines. However, they were mostly new recruits; so new they had never been to sea before and had no idea how to load and fire their muskets.

First port of call on the voyage south was the island of Madeira for much-needed replenishment. The squadron had taken four weeks longer than the usual ten days to reach the Portuguese possession off the coast of North Africa and must have been getting low on rations and water before they tackled the Atlantic proper. The local people told them that a Spanish squadron had been seen lurking in the area. So near to the Iberian Peninsula it's likely the British were seriously outgunned.

They got under way as soon as they could and to make up time the cargo of the transport ship *Industry* was transferred to the decks of the other ships while under way (a procedure known today as "RAS," replenishment at sea). It took three days. Once this difficult task had been completed the *Industry* was allowed to return home, no doubt to the great relief of her crew.

Conditions on Lord Anson's passage southwest across the Atlantic to South America were appalling. In the tropical heat the food started to rot. Nothing is chilled, nothing comes in cans or dried ration-packs. Of course, you eat the salted and cured stuff first but what do you do when that runs out and you're *really* hungry? You eat the meat after scraping the green mold off the surface; you pick the weevils out of the HARD TACK biscuits. The ships became infested with flies, a problem usually solved by opening all the gun ports and letting the wind carry them away. But that wasn't an option available to Anson's ships. They were overloaded, so low in the water that opening the gun ports would have let in more ocean than air.

Conditions were pretty insanitary on all sailing ships. The crews would urinate over the GUNWALES to the leeward side; using the windward side is a mistake you only make once. The other facility was called the THUNDER BOX. This was a man-sized wooden box rigged over the side. It had a hole in the floor. In anything above a Force 7 it could also serve as a heart-stopping bidet.

What made this even worse were the marine *invalids* restricted to their hammocks on the lower gun decks. There was an Admiralty convention stating that hammocks should be hung 14 inches (0.35m) apart. But these ships needed manning around the clock so only half the crew would be below deck at any one time and they would make a sensible arrangement to occupy alternate hammocks. Although this worked well for HMS *Wager*'s normal complement of 140, on this voyage she was playing host to 360 marines. The professional sailors didn't like these passengers on deck because they got in the way of the orderly running of the ship. It was crowded below and many of the *invalids* could not make it up the companionway and over the side into the thunder box.

mark, "wide of the mark"

a long way off the LEADING LINE; in the wrong ballpark, inaccurate, very wrong (vernacular expression).

marks

physical entities on land or at sea used for the purposes of NAVIGATION; marks may be designed for the job (e.g., lights on the end of MOLES marking the entrance to a harbor) or improvised on the grounds of good visibility and fortunate location (e.g., a church spire which, when used with another mark, provides a safe LEADING LINE into the harbor); also the depth indications on a LEAD LINE.

marlin spike

a tool used for working with ropes and shackles; originally made from hardwood, now from steel and extending to a point at one end.

marry

"marrying" the process of interweaving the un-LAYED strands while splicing two ends of rope together.

martingale (UK), vang (US)

any line or device used to prevent a BOOM from lifting; the most common example on a modern yacht is the KICKING STRAP.

Mary Celeste

The strange circumstances of the abandonment of the *Mary Celeste* is rivaled only by the mystery of why Bermuda Triangle buffs should think that the Azores are anywhere near the southwest Atlantic.

The *Mary Celeste* was a BRIGANTINE in passage between New York and Gibraltar in the winter of 1872. For those of you unfamiliar with the basic facts, the story goes like this. At 1400 local time, to the east of the Azores, another brigantine called the *Dei Gratia* sights the *Mary Celeste* ahead of them. She is sailing under two headsails only, so the *Dei Gratia* soon catches up. They hail the other ship, but there is no response. The *Mary Celeste* seems to have been abandoned; her only boat and crew of ten (including the captain's wife and two-year-old daughter) are missing. An inspection by the skipper of the *Dei Gratia* shows that the ship is in bad condition: the hatch covers are off, the MAINSAIL and MIZZEN are blown out, the log is missing, the cargo hold is waterlogged, and the BINNACLE has been knocked off its mounts.

So what happened?

- They encountered a storm, feared they were sinking, and took to the boat. Then the boat sank but the ship kept going. This is not an uncommon occurrence. Or . . .
- In bad weather some of the 1,701 barrels containing a hazardous cargo of denatured alcohol split open and they feared an explosion (these were the days of kerosene lamps with naked flames). So they took to the boat. Then the boat sank. Or . . .
- There was a conspiracy! By a remarkable coincidence, the *Mary Celeste* and the *Dei Gratia* had been moored next to each other in New York. The evening before departure, the skippers of the two ships had dined together. Were they plotting an insurance scam? Was there a cover-up by the US government? Is there mention of the *Mary Celeste* in the FBI's X-Files?

Probably not—they almost certainly met by chance. In 1872, without GPS, VHF, or short-wave radio, it is most unlikely that a planned rendezvous on the high seas would be successful. It is boring, I know, but I vote for the second option as the solution to this not-very-mysterious, but very sad loss of a skipper, his family, and his crew.

mast

a vertical SPAR used to support a sail so that it will catch the wind; a mainmast supports the MAINSAIL, the mizzenmast, the MIZZEN. Modern masts are usually hollow, so that the downfalls of the HALYARDS and TOPPING LIFT can be carried internally. Also used to hold up a whole lot of other stuff; radio and GPS antennae, CARD, RADAR scanners, and WIND VANES. On most rigs, especially BERMUDAN, the mast is held vertically and as rigid as possible through a system of STAYS. The FORESTAY and backstay inhibit fore and aft movement while lateral movement is prevented by the SHROUDS. This arrangement puts considerable stress on the mast and gives thought to the analogy with a bow and arrow (the arrow being the mast pointing at the keel). It also puts considerable stress on the rigging, which is constantly under tension. The consequences of this are (a) reliability—if anything is going to break it will be the mast or the rigging—and (b) cost—25–30 percent of the cost of modern sailboats is in the rigging. Could there be a better way of doing things? The implication of using an *UNSTAYED* mast is that the mast—regardless of whether it is made of wood, metal, or synthetic materials—will need to be flexible. But it is always the bendy trees that survive the storm.

masthead

the very top of a MAST.

mast step

the point at which the very bottom of a MAST is secured; today, a mast is usually stepped at the DECK or at the keel (in which case the lower part of the mast will stand up through the main cabin).

Mayday

radio distress call; from the French *m'aidez*, "help me." See DISTRESS SIGNALS.

McGonagle's Law

MURPHY said that "If anything can go wrong, it will go wrong." McGonagle waved a hand dismissively and declared that Murphy was an incurable optimist.

measurements of boats and ships

The exact measurements of a yacht can be quite important; they can determine the maximum possible speed and, more importantly perhaps, can determine how much one pays for a MARINA berth. The key measurements of are as follows:

<u>Length</u>

- Length overall (LOA); the measurement from the foremost to the aftmost part of the vessel, including everything. But does that include the BOWSPRIT when the bowsprit is retractable? It is possible to have an interesting discussion with a marina manager about that.
- Length on the WATERLINE (LWL); foremost to aftmost of the WATERLINE when the vessel is loaded normally.
- Length on DECK (LOD); foremost to aftmost of the deck.
- Length between perpendiculars (LBP); the distance between the forward part of the stem and the after part of the rudder post.

Width

- BEAM; the width of the vessel, not including the fittings (such as navigation lights) and the RUBBING STRAKE (US: rubrails).
- Extreme beam; the width of the vessel, including the fittings and the rubbing strake (US: rubrails).
- Beam AMIDSHIPS; the width of the vessel when measured exactly midway between the STEM and the STERN.
- Beam at the WATERLINE (BWL); the width of the vessel at the plane of the waterline.

Depth (draught, draft)

- Depth; the vertical distance from the SHEERLINE to the bottom of the KEEL.
- Molded depth; the distance from the DECK to the top of the keel. Only useful, I suspect, if the keel drops off.
- Draft—the vertical measurement from the designed waterline to the lowermost edge of the keel.

Overheard at a boat trade show:

> PROSPECT: How deep is this boat without its keel?
>
> SALESMAN: Hmmm, about forty-five feet, sir.
>
> PROSPECT: Eh? Really? How come?
>
> SALESMAN: Well, she'd be upside down, wouldn't she, sir?

Tonnage

- Displacement tonnage; the weight of water displaced by the hull when floating at its load WATERLINE. Eureka!
- Deadweight tonnage; the total weight of cargo and stores a vessel can carry.
- Gross tonnage; the total internal volume of a vessel; 100 cubic feet is considered to be equal to one ton.
- Net tonnage; the space within a vessel available for cargo, 100 cubic feet is considered to be equal to one ton.

mega yacht

a luxurious motor or sailing yacht, usually much more than 30m (100ft) LOA; see CORINTHIAN.

Meltemi (Etesians)

the predominantly northerly summer prevailing winds in the Aegean Sea; "Etesians" from "etos" is from the Greek and means "annual," but they are most usually referred to by the Turkish name of "Meltemi."

Such are the somewhat-strained diplomatic relations between the Greeks and the Turks that "Etesians" is the preferred expression in a Greek taverna. According to the bible—Imray Laurie's excellent *Greek Waters Pilot*, by Rod Heikell—the Meltemi begins in June and eases off by September, but personal experience suggests that this vicious blow can start as early as April (see RAFTING). Getting a good weather forecast in Greece is important; a DEAD CALM, Hades-hot day can within hours become a struggle to find shelter. The solution is to find a taverna with a television, buy a few beers (locally brewed Carlsberg seems to be the only option) and ask an English-speaking local to translate for you.

Mercator projection

a projection of a map of the world onto a cylinder so that all the parallels of latitude have the same length as the equator. First published in 1569 and used especially for marine charts and certain climatological maps.

meridian

any north–south line of LONGITUDE.

midshipman

Webster's 1828 edition: "n. In ships of war, a kind of naval cadet, whose business is to second the orders of the superior officers and assist in the necessary business of the ship, particularly in managing the sails, that he may be trained to a knowledge of the machinery, discipline and operations of ships of war, and qualified for naval service."

(Refers to the adventures of Commander John Byron. See also GIANTS OF PATAGONIA, MARINES, MUTINY, SHIP FEVER, and SQUADRON.)

The recruitment of boys wanting be naval officers was introduced in Britain in 1660. Their contracts required ship's captains to assume a duty of care and to educate them. The letter of service stated that they should receive "such kindness as you shall judge fit for a gentleman, both in accommodating him in your ship and in furthering his improvement" ["Midshipman," *Encyclopedia Britannica* XVIII (11th edition)].

When, in 1740, the Honorable John Byron sailed from England for the Pacific, he was eighteen years old and already a ten-year veteran of life in the Royal Navy. This voyage was going to be a life-changer, a tale of the ravages of SHIP FEVER and dysentery, shipwreck, MUTINY, a life-or-death struggle to survive as a CASTAWAY in one of the bleakest parts of the world only to be taken prisoner by the enemy.

"Foul-weather Jack," as he became known to his crew, was a member of a seafaring family and the grandfather of George Gordon Byron, Lord Byron, the romantic poet. He'd joined the navy at the age of *eight* and was sent straight off on a circumnavigation. Today that would be hard to comprehend for a parent on the school run and was probably the beginning of the end for adventure holidays. But it was not all that unusual in the 18th century and young Foul-weather Jack survived and developed as a midshipman, wearing the smart uniform of blue tailcoat, white shirt and waistcoat, white breeches and stockings, and polished black shoes. He probably wasn't given a sword until he was tall enough to avoid tripping over the long scabbard as he rushed up and down the COMPANIONWAYS.

For more on the Honorable John Byron, see GIANTS OF PATAGONIA, MUTINY, and SQUADRON.

midships

the FORE-AND-AFT centerline of a boat.

milk

1. milking is the process of easing the SHEATH of a rope over a SPLICE.
2. the stuff that has turned into penicillin even before you have rounded the MOLE.

miss stays

if, in attempting to GO ABOUT, a yacht fails and flops back onto the original TACK, she has "missed stays."

> On Friday morning at day light made sail round the point for the Harbour Owharre where we anchored at 9 o'Clock in 24 fathom water, as the wind blew out of the Harbour I choose to turn in by the Southern Channell, the Resolution turn'd in very well, but the Adventure *missing stays* got a shore on the reef on the north side of the Channell.
>
> *The Journal of Captain Cook*, THURSDAY, 2 SEPTEMBER 1773

Author's emphasis.

mizzen (or mizen)

a small sail mounted on a mizzen mast at the STERN of a yacht; a yacht with a mizzen is called a KETCH or a YAWL (where the mizzen is stepped aft of the RUDDER POST).

mizzen mast

a mast at the STERN of a vessel to which a MIZZEN sail is attached.

mizzen staysail

a triangular sail set forward of the MIZZEN MAST.

monkey's fist

a means whereby a rope's end is made into a sphere or ball of rope to enable the rope to be thrown (e.g., ashore or to another ship).

monohull

a boat with a single hull; see also CATAMARAN, TRIMARAN.

monsoon

a heavy seasonal rain (May to October) experienced throughout Southeast Asia and all the way north through China and Japan. Effectively caused by the Mother of All Sea Breezes in areas and times of high humidity. Heated air rises in a convection current over land, drawing in moist air, lifting that, and causing it to turn into serious rain. See BREEZES.

s/v Monsoon

This is the first of three entries that tell the story of a 1930s passage from Denmark to the Pacific Ocean. Part 2 is under GALAPAGOS ISLANDS and Part 3 under VANIKORO SHIPWRECK.

The idea for the now long-forgotten voyage of the SCHOONER *Monsoon* started in the head of a Danish schoolteacher Alex Möller in the early 1930s. Möller, the headmaster and owner of Vejlby Village School near Aarhus, organized a meeting of those friends and acquaintances he thought might be interested in sailing with him on an expedition to the South Seas. The reaction was mixed but generally positive. Then, to prove his seriousness, Möller sold the school. One of his friends, Hakon Mielche, said that he was "caught by both feet in the birdlime." No, I can't imagine what he meant, either, but in a book he wrote about the adventure, he took up the story thus:

> The avalanche gained momentum. The plan took shape. The ivory was jettisoned owing to a certain ignorance of the dental mysteries of elephants and the market for this raw material for piano keys and toothpicks, but the idea of the South Seas was planted more and more firmly in Headmaster Möller's head. He wanted to get out into the world and have a look around.

There being a distinct shortage of elephants on Nukunono, jettisoning the ivory (the group's original source of revenue the trip) was a really smart move. But they were not short of ideas.

> How far could we pay for a pleasure cruise round the world by purchasing ethnographic rarities which could be sold to museums and private collectors? The Etnografisk Samling [Ethnographic Museum] was, it turned out, interested in the affair, and the inspector gave us several good tips. Later the Zoological Museum came along and its director declared that he would welcome the opportunity of extending the museum's collection with a comprehensive exhibit of the fauna of the South Sea Islands.

Was the fauna to include one of the famed elephants of Bora Bora? A live elephant would demand a big boat.

> One day Möller was strolling along the bank of the Frederiksholm Canal in Copenhagen—and he came across the Monsoon. It was a case of love at first sight. She was built forty years ago in Boulonge-sur-Mer from good French oak, and furnished with pitchpine masts made to withstand any storm—which was necessary, too, for she was intended for fishing in the North Atlantic. Although she had led a restless existence since then and had been put to many other uses, she was still completely seaworthy. . . .
>
> Then Mr Möller bought her. If ships had tails to wag, she would have wagged hers. She was just the ship for which he had been looking for so long—it was almost as if she had been built for the purpose. Numerous borings and examinations from stem to stern, from the top of the mast to the keel, showed clearly enough that there was no possible doubt as to her seaworthiness. She was towed to Erichsen & Grön's yard, and Mr Möller, taking up residence with his family in the cabin, personally superintended the process of rejuvenation.
>
> As the refitting progressed, Möller made a start on begging, borrowing or even buying provisions.

The results piled up on the Monsoon's deck. Guns, revolvers, nautical instruments, fuel oil, charts, tarpaulins and a thousand other things were willingly lent us by the parsimonious Naval Stores in such generous quantities that we began to be seriously afraid of what would happen should war break out during our absence. A chemical factory sent us vitamins in bottle and pills, both A and B plus the whole alphabet, sufficient to last for many years; while the gifts of friends and equipment we had borrowed flowed in over the rail and filled the hold up to the brim.

A paid crew of five was taken on. They were as character-full as the Monsoon herself.

Peter Bundaberg Thomsen became skipper. He was engaged on the warmest of recommendations from Knud Andersen and because he was one of those real good old windjammer men whom it hurts when they see as much as the smoking chimney of one of those modern coal-buckets described in the dictionary as "steamers." He had sailed the seven seas, chewed salt horse on his way round the Horn and shortened many a sail in many a squall on the Tasman Sea. . . . He sang chanteys, but would have done better not to. His memory was such that he was excellent at forgetting what time it was twenty-four hours ago, or where he had put his foul but beloved pipe; but ask him the name of the ship that passed Cape Hatteras at four o'clock in the afternoon of 4th April, 1898, and, without blinking, he would tell you that it was the Danish barque Margaret, that she had sprung a leak on the starboard waterline three feet below the Plimsoll line, that she was carrying that-and-that sail, that the chronometer lost two seconds a week, and the wind was three points aft, strength seven, and that the master was called Olsen and drank like a fish.

Aquavit, no doubt. Soon the day approached for Monsoon to set sail.

The Monsoon came to life. There were sounds of whistling, hammering, cleaning and carpentering, and then came the awful day when the 35 h.p. motor had to be shipped. That was a bit awkward. [It's always a bit awkward!] It was something of a sacrilege—like putting steel furniture in a rococo drawing-room. However, we had no choice. If we didn't want to row Monsoon half the way, the 35 h.p. would have to come to our help, for our voyage took us suspiciously near the region of calms round the equator where the wind is one day as incalculable as a woman—according to the classics—and the next week disappears altogether and goes to sleep on the horizon under the clouds drifting with the trades, and laughs at the poor hulk with its fluttering sails and cursing crew. . . .

> Dr. Mortensen, an old East Indiaman made a short farewell speech and gave us a word or two of good advice for the road. It was the first time that the idea of malaria or head hunters occurred to us properly, and the first time that the pleasant shiver of anticipation ran down out spines. Then came good wishes, port wine, biscuits, emotional returns of thanks by Mr Möller, curious lumps in one's throat, rapid steps up the companionway and over the bridge. The deck was cleared, the hawsers drawn in—a bevy of white handkerchiefs—faces which became more and more indistinct. The first waves in the bay began to splash against the bows: the sails came to life and filled; the voyage had begun.
>
> Excerpts from *Let's See If the World Is Round* by Hakon Mielche, 1944 (originally published in Danish, 1938, now sadly out of print)

The voyage was to take them via the Salvage Islands and Tenerife, across the Atlantic to St. Thomas before passing through the Panama Canal. Then Pearl Island, the Galapagos, the Marquesas, Tahiti, Samoa, Fiji, New Caledonia. . . . To read more about the voyage of the Monsoon, see GALAPAGOS ISLANDS and VANIKORO SHIPWRECK.

mooring

effectively a permanent ANCHORing place in which the end of the RODE is attached to a float and the "anchor" may be a big lump of concrete or even a large anchor. Moorings may be privately owned or the property of a yacht club or port authority.

A mooring is always preferable to using your own ANCHOR, especially where the holding characteristics of the bottom are unfamiliar. The floating BUOY attached to the rode usually has a ring or a handle on top. The pick-up routine involves having someone in the BOW with a BOATHOOK. The helm then steers toward the mooring, against wind and tide as far as possible.

The closer the yacht gets, the harder it will be for the helm to see the buoy. At this stage, the person in the bow will need to employ hand signals until the buoy is within boathook range. The buoy is then deftly lifted over the PULPIT and its line used to pull the rode proper in far enough so that it can be passed over the bow roller (under the pulpit) and secured to the BITTS, a sturdy CLEAT, or whatever. Great satisfaction accrues from doing this with style on the first pass with bonus points for doing it all under sail.

If you cannot get a heavy buoy onboard the trick is to pass a line through the ring or handle and secure that. The downside to this is that the buoy might start to chafe against the HULL when the wind shifts. This can keep you awake—worrying about how much damage is being done to the paintwork. If your boat has a BOWSPRIT, you have a solution to this problem; secure the buoy to the outer end of the bowsprit using a BULL ROPE. By the way, do try and ensure that you are using a *visitor's* mooring; otherwise the owner might arrive back in the middle of the night and rightly demand that you move.

The Greek island of Santorini (Thira or Psira) presents an interesting challenge for yachts visiting the tiny harbor of Thira town. The island includes a lagoon formed in a volcanic crater. The surrounding craggy cliffs soar to 300m—and down 300m, too. Just a few boat-lengths off the pier, the depth is already over 20m and the bottom less than ideal for anchoring. So, if you want to tie up stern-to as local tradition dictates, you need to make use of one of the permanent MOORINGS. These, however, are not orange basketballs bobbing on the surface; they are the kind of huge rusting metal drums (2m in diameter) that supertankers tie up to. Having a close encounter with one of these mooring would seriously compromise your paintwork.

The solution is to blow up the inflatable dinghy and get out the oars. The fittest crew member is then rowed out with a line. He leaps onto the BUOY (no mean feat from an inflatable) and secures the line through the ring. By the time the dinghy crew get back, the beers have been broken out and planning starts for the journey up the cliff (by donkey or funicular railway) for dinner and the amazing view over the supposed site of the Atlantis legend.

mooring single-handed

Solo sailing is at its most hazardous when one is in an inshore or pilotage situation. It is at its funniest when trying to pick up a MOORING. There is expensive

hardware to bump into and folks around to laugh when you get it wrong. You really need to stay in the cockpit to control the boat. So how are you going to pick up the marker **BUOY**? This method works (with practice):

- Permanently fix a single-sheath **BLOCK** to one side of the **BOW ROLLER**.
- Run a line through the block. The rope needs to have a **SNAP-SHACKLE** at one end. This end runs along the hull outside of the stanchions, through the block. The other end runs back along the side-deck inside the **STANCHIONS** to the cockpit. The shackle needs to be clipped on somewhere convenient alongside the cockpit.
- Sail toward the buoy. Correction—motor toward the buoy. As you get closer, maneuver so that the buoy comes down the side of the hull where you have rigged the line.
- Ease off the sheets (or go into neutral).
- Lean over the side and grab the buoy, quickly snapping the shackle onto it.
- Now release the buoy and start to haul on the other end of the line. Take your time. The boat will move into whatever position is dictated by wind and tide as the buoy comes around to the bow. Tie off the rope and make your way forward to pull the buoy aboard, secure properly, and enjoy the applause of your watching neighbors.

Practice without a crowd until you can do this with style and confidence. If your **FREEBOARD** is high or you are having trouble reaching the buoy with a boathook, try this arrangement.

- Obtain an aluminum or strong GRP tube 1.5 to 2m in length.
- Run a suitable rope though the tube and loop the end back.
- Fix about 10–15cm of the line to the tube, starting 15–20cm from the end of the tube. Do this using a strong adhesive and then whipping.
- You now have a lasso with a rigid handle. The size of the loop can be adjusted by pushing or pulling the rope through the tube.

Now, as the buoy comes down past the **COCKPIT**, hold out the “lasso” with one hand, passing it over the whole buoy. As you catch it, pull the rope through the handle until the loop is tight. Drop the handle and haul in using the rope.

This is a handy multipurpose gadget that can be used for retrieving objects (including crew members) from the water. When you catch that $83,000 **TUNA**, you don’t want to damage it, so lasso its tail and haul it over the side using the main boom and the topping lift.

mosquito

malaria is still a major health problem—the contaminated eggs of the mosquito kill three million people a year worldwide. Endemic in tropical Africa, most of Asia, the Caribbean, South and Central America, and the Pacific Islands. Just about anywhere you want to sail, really.

If heading for a malarial zone, check before you set sail the recommended medication (medical centers have access to databases) and take more than enough of it with you for the whole crew. It might not be readily available at your destination and it only takes one bite while unprotected to put you at risk from this debilitating and deadly disease. The instructions on the bottle will tell you to start taking the pills some weeks before entering a malarial zone and for some weeks after leaving: be disciplined about this.

And if the malaria doesn't get you . . . the mosquito also transmits yellow fever, dengue fever, elephantiasis, Japanese encephalitis, and a bilge-load of other nasty conditions. All can be fatal; yellow fever and Japanese encephalitis are incurable.

I confess to having a bit of a cavalier attitude to the "mossie." They don't seem to fancy me as a breeding ground. But I have worked with people who have been stricken with occasional bouts of malaria and that has always encouraged me to keep taking the tablets. Political correctness is not on the agenda for the mosquito. Kill them. You will never get them all, so do what you can to stop them biting you.

Protecting a boat from mosquitoes is not simple and needs to be undertaken before you get to a malarial region. Chandlers do not always stock the mosquito netting you are going to need, but upmarket camping shops always seem to stock it. PORT-LIGHTS, windows, DECK hatches, and the COMPANIONWAY hatch are all likely to be left open in the heat of a tropical mooring. Use your ingenuity (and the mosquito netting) to fashion insect screens for all these openings to the accommodation. If you are likely to sleep on DECK or in a HAMMOCK under the stars, you will be able to rig a net for that, too. Nets should be impregnated with an insecticide called "Permethrin." This can be bought as a spray so that nets can be retreated from time to time.

Tip 1: Make sure there is clearance between your body and the net. If the net touches your skin, the mossie will be able to bite you through it.

Tip 2: Deploy the anchor light (riding light). The COLREGS (US: NAVRULES) require you to do this any, but the light will draw many flying critters away from the temptations of the flesh.

Sadly, none of the above will have much effect against the Aedes mosquito, which spread dengue fever and yellow fever—they only bite during the day. (A few years ago a Norwegian friend of mine contracted dengue fever in Delhi of all places. It damn near killed him.)

By the time you put the screens up, your boat will already have insect stowaways. Use a DEET-based insect repellent to deal with these. Burning special incense coils may also help.

Mother Carey's Chickens

The Stormy PETREL. From the Latin *Mata Cara*, "Dear Mother." The French call them *Oiseaux de Notre Dame* (small bird, big name). Sailors also used to call snow "Mother Carey's Chickens."

Mother Carey's Goose

the Great Black Petrel or Fulmar.

motorsailer

a hybrid cruising vessel that has sails and engines more powerful than would be needed for berthing and battery charging.

motorsailing

sailing with the engine running and in gear at the same time; COLREGS require an inverted black cone to be carried forward of the mast when motorsailing, but rarely do you see them.

mouse

to wrap twine or wire around any device (such as a shackle) to prevent it from opening accidentally.

multihull

a boat with more than one hull; see CATAMARAN and TRIMARAN.

Murder on the High Seas

Well, it was just off the coast of Devon in the English Channel, actually. What made this murder particularly interesting is that it might have been the first-ever case where a GPS receiver "appeared" as a witness in a criminal trial.

Late in July 1996 a Brixham fisherman called John Copik hauled in his trawl net and found he had a very interesting catch. It was a hefty ANCHOR in good condition and with a short length of chain still attached. He already knew someone who was in need of an anchor. Also in the net, coincidentally, was the dead body of businessman Ronald Platt. Copik handed that over to the police but kept the CQR.

By June 1998 Canadian fraudster Albert Walker was on trial at Plymouth Crown Court for the murder of Platt. Local detectives alleged that Walker had taken his former business associate sailing on his 27ft yacht *Lady Jane*. The Canadian had murdered Platt, tied the anchor to him, and thrown his body over the side. By the time Copik had trawled up the body (two days later) the anchor had become detached and he had been lucky to get it in the same net.

But the clincher for the prosecution was the GPS receiver. When this was discovered on the yacht by the police, they had it examined by the manufacturer. They were able to tell the police exactly when the receiver had been switched off and the latitude and longitude it was reading at the time. This turned out to be 20 July 1996—two days before the discovery of Platt's body—and the position was right where it needed to be to drift with the intervening tides toward Copik's waiting net.

Movie screenwriters were sitting in court to hear the jury's verdict of "Guilty." Albert Walker was sentenced to life imprisonment and the BBC made the well-reviewed film.

Murphy's Law

If something can go wrong, it will go wrong. Never more true than on a boat. But the traditional US seafarer's expression is "Shit happens!" See also MCGONAGLE'S LAW.

mutiny

rebellion against established authority; on a ship, a revolt by sailors against the officers. (Refers to the adventures of Commander John Byron. See also GIANTS OF PATAGONIA, MARINES, MIDSHIPMAN, SHIP FEVER, and SQUADRON.)

The first officer to die onboard Lord Anson's 1740 expeditionary SQUADRON to South America (see SEA FEVER) was the splendidly named Dandy Kidd, the Honorable John Byron's captain onboard the ominously named HMS *Wager*. His death was a bad omen. Kidd had to be replaced by the relatively inexperienced Lieutenant David Cheap. Rather than swagger around the quarterdeck surveying his new domain, Cheap took to his bed with what seems to have been a case of chronic *mal de mer*. (See SEASICKNESS.) Junior officers had to assume responsibility for the day-to-day business of the *Wager*.

As the squadron rounded Cape Horn in a full gale the eight vessels became separated. No longer could they tiptoe along behind the commander's ship; they had to make their own navigation decisions and try to reach the rendezvous point at Juan Fernández Island. On 12 May 1741, David Cheap—now Captain Cheap—got a key decision wrong. Instead of sailing well clear of Tierra del Fuego west into the southern Pacific, he made the turn north too soon. After a few days they realized they were in a bay and on a lee shore from strong westerly winds. Cheap ordered the helmsman to turn the ship around and they added more sail, desperately trying to get into open waters again. The crew must have been terrified as they battled against the weather throughout the night. Sixteen hours later, with Captain Cheap below decks with a dislocated shoulder, they hit the rocks and started taking on water. It was pitch dark. Sick and injured sailors were drowned in their bunks. The fierce wind continued to hammer into the rigging, making the decks shudder alarmingly. The bell clanged. The steering system broke and they could only maneuver the *Wager* using the sails.

Then all hell broke loose.

A large group of exhausted, terrified sailors lost their nerve. They went below and helped themselves to the rum barrels in the store. They dressed in officer's uniforms and took muskets from the armory the *Wager* was carrying for the squadron's shore parties.

Wisely, the officers and the rest of the crew ignored them, launched the boats, and rowed ashore safely. The following day they watched as the hull of the ship split along her bilge and flooded, drowning the drunken sailors left onboard.

As HMS *Wager* disappeared beneath the waves everything changed for the 140 survivors standing on the beach. The first change was a legal one: with the wrecking of the ship, all the officers lost their commissions. Newly promoted David Cheap was no longer the captain; he wasn't even a lieutenant any more. He was now *Mr.* David Cheap. It wasn't quite so bad for the eighteen-year-old midshipman John Byron; at least he was the Honorable John Byron again, but that didn't give him any authority over the other members of the crew. They were all in it together now.

Byron didn't know it at the time but they were on an island (today called Wager Island) that is part of an archipelago on the southern side of the Gulf of Penas at the bottom end of Chile. "Pena" is Spanish for sadness or embarrassment and there were plenty of both around in May of 1741. Any attempt by the former officers to demonstrate their leadership skills was probably doomed to failure. At the same instant the officers lost their status the crew lost their pay. If they'd had any pension rights back in the 18th century they would in all likelihood have lost those too.

When they should have been pulling together on this bleak island to make shelter and find food (easy for me to write that) the ex-officers and ex-navy ratings started to fight among themselves. It wasn't the kind of brawl these tough men might have experienced outside the Sally Port Inn back in Portsmouth; they were murdering each other. The chaos was sparked off by David Cheap (his seasickness miraculously cured) when he shot and killed one of the crew. The former captain became a marked man and took to carrying two loaded muskets in his belt. The incident became known as "the *Wager* Mutiny" but I'm sure a good defense lawyer would point out that the sailors no longer had any officers of standing to mutiny against.

Eventually eighty of the seamen took to the boats and headed south, planning to round Cape Horn again, reversing their tracks to make their way north through the Atlantic. Only five from this group reached England again but it was a remarkable and brave demonstration of seamanship. John Byron was among twenty men who stayed put on Wager Island, living off scraps of rations from the wrecked ship. They were not alone; Indian fishermen had seen a business opportunity and were waiting. After striking a deal for food and weapons, the local people took the Englishmen and Irishmen north in their canoes to Valparaíso where they were handed over to the Spanish.

Byron and the others were lucky. When the main body of the squadron had raided the port months earlier, Anson's officers had treated their Spanish prisoners with decency before releasing them unharmed. That respect was reciprocated by the garrison commander and, after being detained for a while, the men of the *Wager* were released on bail—"port arrest" rather than "house arrest." Their captors knew they weren't going anywhere fast. In fact they were there for nearly four years. It seems that, after Wager Island, this was no great hardship. Byron, the handsome young English gentleman, became a favorite of Valparaíso's *señoritas* and probably improved his Spanish as a result.

In the meantime, Commander Anson and the rest of the squadron were back in England. Since parting company with the *Wager* he had visited (or raided) Concepción, Valparaíso, the Island of Juan Fernández, Paita on the coast of what is now Peru, Quoibo off Panama, Acapulco, the Marianas right across the Pacific, Macao in China, a side-trip to the Philippines, the Strait of Sundra, Cape Town, and then back to Portsmouth, arriving on 15 June 1744. Of the 1,900 men who had boarded the squadron's ships in 1854, only 188 made it home. The majority of those lost had died of malnutrition or disease.

Midshipman John Byron and Captain David Cheap got back to England from Santiago in June 1746. There had been 296 men onboard the *Wager* when she went aground; only nine survived. Twenty-seven years later Byron wrote a book about his experiences on the fated voyage. It was a bestseller, being printed in some nine editions. (John Byron: *Narrative of the Hon. John Byron; Being an Account of the Shipwreck of The Wager; and the Subsequent Adventures of Her Crew*, 1768.)

The rewards were much greater for George Anson. His captain's pay for the voyage amounted to £719. However, he was entitled to the biggest slice of the prize money for the Spanish treasure galleon *Nuestra Señora de Covadonga* seized off Manila. That reward totaled £91,000, the equivalent of £8 million in today's money.

For more about the Honorable John Byron, see GIANTS OF PATAGONIA, MIDSHIPMAN, and SQUADRON.

nappy (UK)

see DIAPER.

National Hurricane Center (NHC: US)

The National Hurricane Center (NHC), part of NOAA, maintains a continuous watch on tropical cyclones over the Atlantic, Caribbean, Gulf of Mexico, and the eastern Pacific from 15 May through November 30. According to its website:

> The Center prepares and distributes hurricane watches and warnings for the general public, and also prepares and distributes marine and military advisories for other users. . . . NHC also conducts applied research to evaluate and improve hurricane forecasting techniques, and is involved in public awareness programs.

This hurricane data is also available via WEATHER-FAX, of course.

National Oceanographic and Atmospheric Administration (NOAA: US)

An administration within the US Department of Commerce responsible for the distribution of nautical charts and weather forecasts. A much bigger organization than the UK HYDROGRAPHIC OFFICE. For anyone passage making in the Atlantic, the NOAA's NATIONAL HURRICANE CENTER provides an invaluable service.

nautical mile (nm)

one minute of LATITUDE; the basic measure of NAVIGATION and, therefore, cannot be replaced by the kilometer. At 1,852m (6,076.12ft), 1nm is about 15 percent longer than 1 statute mile.

navel pipe

a pipe that leads the ANCHOR chain from the FOREDECK below to the CHAIN LOCKER.

navigation

the honorable skill of being able to (a) determine, at any instant in time, using whatever means available, the location of a vessel at sea to the greatest degree of accuracy possible and (b) plan the safest and most expeditious route between any points of departure and arrival. This is sometimes claimed to be an art.

> The art of navigation demonstrateth how, by the shortest good way, by the aptest direction, and in the shortest time, a sufficient ship,

> between any two places (in passage navigable) assigned, may be conducted: and in all storms and natural disturbances chancing, how to use the best possible means whereby to recover the place first assigned.
>
> Dr. John Dee, 1570

Two hundred years later there came an injection of stark reality.

> It is far better to have absolutely no idea of where one is—and to know it—than to believe confidently that one is where one is not.
>
> Jean Dominique Cassini (Director, *Observatoire de Paris*, 1770)

I know of people who claim that the ability to operate a GPS receiver has replaced the ability to navigate. This is like saying that an ability to play a CD makes you a musician. The fine craft of navigation did not die with the invention of the sextant; nor did it die with the introduction of radio aids such as Decca and Loran. And it won't die with the advent of the Global Positioning System.

See GPS.

navigational aid

any instrument such as a sextant, dividers, plotter, parallel rules, charts, tables, and methodologies used in practical navigation onboard a vessel. Not to be confused with an AID TO NAVIGATION; this describes lights, buoys, and other external devices or structures. So, a navigation satellite is an aid to navigation; but a GPS receiver is a navigational aid.

The strangest of all would-be navigational aids was the proposed use of the ship's dog to determine LONGITUDE. A close second must be GALILEO's *celatone*, a brass helmet with a telescope fitted over one eye. At a time when few scientists believed that the spring-wound clock would ever be accurate enough for navigation purposes, the smart money was going on a French-backed system based on the observation of the eclipses of Jupiter's moons. The practical problem for seafarers was that you need a telescope to see the planet's satellites and telescopes and heaving decks don't go well together. Galileo's helmet enabled the navigator to locate Jupiter with one eye while the other observed the moons through the telescope. The *celatone* must have looked like one of those high-tech gun sights used by the pilots of Apache helicopter gunships and I can't imagine skipping around the deck while wearing one. The underlying method was sound and very accurate; Captain James Cook used it to calculate longitude during his first voyage to the Pacific in 1768–1771 but I assume that most of his sights were performed on the beach of each new island he discovered. However, as soon as he could get his hands on one of the new-fangled H4 chronometers made by John Harrison, he switched allegiances.

navigation hazard

see MARIGOT BAY.

Navigation Rules (US)

The "NavRules" are a set of regulations governing the "rules of the road" for vessels at sea; essentially the same as the COLREGS for the high seas but also contains the rules for navigation on US inland waters.

navigation table

an area incorporating the chart table, instruments display, electrical supply breakers, and storage space for charts and instruments; may also include a seat although some navigators, like Ernest Hemingway, prefer to work standing up. Also called "the office."

Navtex

a regionally based text radio broadcast service providing a continuous stream of weather and navigation data to both commercial shipping and recreation seafarers. Received text can be printed on something like the roll of a supermarket check-out till or shown on an LCD display. Programs are available for laptop computers hooked up to short-wave radios so that the text can be displayed on the screen. A very useful, free service very popular in European waters although I have never seen anyone using it in US waters.

neap tide

a TIDE with the smallest range between high and low water.

night vision

the ability of the human eye to adjust to low levels of light is often referred to as "night vision." The eye achieves this by opening the iris as widely as possible; in higher light levels, the iris will close—but both these changes take time to come fully into effect. For the helmsman in particular, it is important for night vision to be protected. One precaution that should always be taken is to reduce the light level in the cockpit instrument display. With NMEA-networked instruments, changing the level of one instrument will cause them all to change to the same level. See NMEA.

A particular problem occurs when moving between the cockpit and main cabin at night. One solution to this is to use dimmed red-spectrum lights at the navigation station, thus minimizing the extent to which the iris needs to change. If this arrangement does not apply, an expedient solution is to close one eye before going into the lighted cabin and keeping it shut until on deck again; fifty percent night vision is better than none.

Niño, Niña

see el Niño.

nm

see nautical miles.

NMEA

the North American Marine Engineering Association. This trade body has set some important standards, the most important of which specifies the way in which instruments can share information over a network. This enables, for example, the wind display to obtain speed-through-water from the log and deduce the true wind as well as apparent wind.

NOAA (US)

see National Oceanographic and Atmospheric Administration.

nocturnal

a nocturnal is an old instrument that resembles an astrolabe but is used primarily for obtaining the time at night. Overlapped discs are graduated for the date and time; when a hole at the center is aligned with Polaris, a co-centric arm can be set against circumpolar stars. The time can then be read off the scale. Before the introduction of clocks that would work reliably onboard ships at sea, the nocturnal was the only means—subject to the night sky being clear of cloud—of obtaining a local time check (accurate to 15–20 minutes). Once a nocturnal time had been obtained, the ship's hourglasses would be set to provide times during the following day or days for watch changes.

nor'wester

any wind originating from the northwest

no-see-um

ceratopogonidae; sometimes called "the biting midge" or "punkey," this flying insect is, on its own, almost invisible to the naked eye. However, no-see-ums are gang-bangers. Worse than that, they are *vampire gang-bangers*. My first encounter with them was shortly after dropping anchor off a small island. The sails were down, the engine off and I was sitting in the cockpit with a Haitian rum, checking out Orion. Looking down instead of up I realized that my legs had turned black. No-see-ums! A whole cloud of them had made their way over the water to gorge on the blood of the biggest mammal around. It was too late for the DEET. (Not to be confused with the sand fly.)

null, null-point

when taking a bearing on a RADIO BEACON, the point at which the signal is silent is called the "null point."

oar

1. a wooden pole with a flat end (the blade) used for propelling a boat. The basic physics of leverage apply; when the oar is still out of the water the fulcrum is about halfway along the oar at the ROWLOCK. When the blade is dropped into the water the fulcrum magically moves from the rowlock to the blade. As the inboard end of the oar is pulled or pushed, the boat moves in the opposite direction to that in which energy is being applied. A small boat such as a DINGHY or TENDER will have one oarsman rowing with a pair of oars, one extending from each side of the boat. Larger boats may have two or more oarsmen. Once the BEAM of the boat gets beyond a certain width, the rowers will sit closer to either side and pull on one oar only.
2. "to put/stick one's oar in": to take part, to get involved, to interfere.

observer

term used in ASTRO-NAVIGATION to refer to a person on a vessel (and their position) attempting to determine their location on the face of the earth (usually through the use of a SEXTANT).

occulting light

a navigation light that is "on" in its reference state but is switched "off" in a predetermined pattern as an aid to identification (compared with the more common flashing light that is “off” in its reference state but that flashes “on” in a predetermined pattern as an aid to identification). Indicated by “Oc” on charts. “Occult” refers to something that is hidden and is from the Latin *occultus*, concealed.

See LIGHT CHARACTERISTICS.

ocean gypsies

people who have abandoned conventional western lifestyles in favor of living on boats and traveling the world by sea. Such people have probably existed since the invention of the boat, but there seems to have been a remarkable increase in this ever-shifting community over the last few decades. See ODYSSEUS.

Odysseus

Odysseus was the “ingenious hero” hero of Greek myth who, after the sacking of Troy spent the next ten years cruising around the Greek Islands as a live-aboard. Now I cannot take issue with Odysseus for this, but while he was battling with Ray Harryhausen–designed monsters in Greek tavernas, his neighbors and business rivals were moving in on the family estate, in particular, his wife Penelope. The Odyssey starts with his arrival back in Ithaca. “Hi Honey, I’m home!”

> TELL ME, O MUSE, of that ingenious hero who travelled far and wide after he had sacked the famous town of Troy. Many cities did he visit, and many were the nations with whose manners and customs he was acquainted; moreover he suffered much by sea while trying to save his own life and bring his men safely home; but do what he might he could not save his men, for they perished through their own sheer folly in eating the cattle of the Sun-god Hyperion; so the god prevented them from ever reaching home. Tell me, too, about all these things, O daughter of Jove, from whatsoever source you may know them.
>
> So now all who escaped death in battle or by shipwreck had got safely home except Ulysses, and he, though he was longing to return to his wife and country, was detained by the goddess Calypso, who had got him into a large cave and wanted to marry him. But as years went by, there came a time when the gods settled that he should go back to Ithaca; even then, however, when he was among his own people, his troubles were not yet over; nevertheless all the gods had now begun to pity him except Neptune, who still persecuted him without ceasing and would not let him get home.
>
> *The Odyssey* by Homer (800 BC), trans. Samuel Butler

According to this account, the Sun-god Hyperion was responsible for Mad Cow Disease, not the British Ministry of Agriculture.

office, the

nickname for the NAVIGATION TABLE and the area surrounding it.

offing

the distance of a vessel from shore; always important to have lots of this—especially with a LEE SHORE. If you are heading into a HAVEN then your offing will be reducing all the time; but that's fine because you are looking forward to stepping onto dry land. Modern vernacular: something is "in the offing" when it is about to happen.

offshore

away from the shore; in the direction of open water.

offshore navigation

navigation that takes place out of sight of land; implies the application of a different set of techniques from coastal navigation, but that has been less true since the advent of satellite navigation.

See GPS.

off the wind

sailing on a BEAM REACH or a BROAD REACH; not CLOSE-HAULED.

oilskins (UK: "oilies")

waterproof, heavy-weather sailing clothes; usually known in the US as "FOUL-WEATHER GEAR" or "FOULIES."

oil spillage

the accidental or deliberate spilling of crude or refined oil into the sea.

Oil spillages are not merely an un-neighborly habit of big oil companies or bulk carriers washing out their tanks. Oil can easily accumulate in the bilges of the smallest yacht with an auxiliary engine. Before pumping out the BILGES, consider the following words of caution from the US Coast Guard:

> The Federal Water Pollution Control Act prohibits the discharge of oil or oily waste into or upon the navigable waters and contiguous

> zone of the United States if such discharge causes a film or sheen upon or discoloration of, the surface of the water, or causes a sludge or emulsion beneath the surface of the water. Violators are subject to a penalty of $5000.

So, apart from being un-neighborly, it can be illegal and expensive.

Tip: Sometimes, if you are taking on water, you have no choice but to start pumping. It is a good idea, therefore, to keep ahead of the oil problem.

- Since prevention is always better than cure, try and stop the oil getting into the bilges in the first place. Use a good-sized funnel when topping up and if you have a spare pair of hands, make sure they are standing by with absorbent kitchen roll to mop up the drips. When changing an oil filter, arrange a waterproof plastic bag under the filter bowl to catch the spill. You can dump the filter in there too.
- Once the damage is done, you may need to clean out the bilges by hand. Start by breaking up the oil using one of the new, powerful washing-up liquids (yes, the same stuff you are using for the dishes and as a salt-water soap). Absorbent paper towels come a bit expensive for this job, but sheets of newsprint left to their own devices for a while will soak it up. Then transfer the paper into plastic trash bags for disposal at an approved site on shore.

A yucky job, but at least you should not have to do it every day.

Old Admiralty (The)

London's Old Admiralty, known as the Ripley Building, still exists. It is the brick-built, U-shaped, three-story building toward the northern end of Whitehall and on the same side as Downing Street. It is reputed to be the first purpose-built office building in Britain.

old salt

an old and/or experienced sailor; usually seen sitting in the corner of the bar.

onboard

on the vessel; to "take onboard" is to understand, to remember in everyday English.

onshore

in the direction of land.

on the wind

sailing CLOSE-HAULED.

opening

when moving away from the line extended from a pair of LEADING or CLEARING marks, the marks will appear to become further apart and are then said to be "opening"; an important concept of PILOTAGE. See CLOSING.

orca

the "killer whale," the one with the bold black-and-white paint job that snacks on seals and is really a big dolphin with attitude and not a whale at all.

outboard

a marine power unit incorporating engine, gearbox, and propeller designed to be mounted on the STERN of a boat. Steering is achieved using a combined handle and throttle on smaller units or mechanical or hydraulic controls on larger ones. Multiple outboard engines can be fitted and I have seen three 200hp units on a single slimline boat—but only Bad People could want something like that!

outhaul

a line that serves to pull the FOOT of a sail tight along a SPAR.

outrigger

outriggers are long poles (usually made from fiberglass) that spread out fishing lines, keeping them well clear of the boat and from each other in order to prevent them from tangling when TROLLING.

overboard

over the gunn'ls and into the water; see LIFELINES, HARNESSES.

overhang

the part of a vessel's hull that extends beyond the WATERLINE at the BOW or STERN.

overtaking

when one vessel overtakes another, COLREGS (US: NavRules) International Rule 13 applies:

a) Notwithstanding anything contained in the Rules of Part B, Sections I and II any vessel overtaking any other shall keep out of the way of the vessel being overtaken.
b) A vessel shall be deemed to be overtaking when coming up with another vessel from a direction more than 22.5 degrees abaft her beam, that is, in such a position with reference to the vessel she is overtaking, that at night she would be able to see only the stern light of that vessel but neither of her sidelights.
c) When a vessel is in any doubt as to whether she is overtaking another, she shall assume that this is the case and act accordingly.
d) Any subsequent alteration of the bearing between the two vessels shall not make the overtaking vessel a crossing vessel within the meaning of these Rules or relieve her of the duty of keeping clear of the overtaken vessel until she is finally past and clear.

No mention of a huge wake rocking a smaller overtaken vessel here, note.

P

painter

a short length of rope used to secure a boat to a jetty or pier.

palm

or "sailor's palm": a leather strip held across the palm of the hand and incorporating a flat metal thimble at the base of the thumb; used by sailmakers and anyone else having to push needles through heavy canvas and sailcloth.

pan-pan

a radio call for non-emergency assistance is prefixed by "pan-pan" repeated three times; such a call might be made in the case of engine failure where the vessel and crew are not in risk of their lives. See also MAYDAY.

pan-pan medico

a radio message prefixed with "pan-pan medico" three times is a call for urgent medical *advice*; the agency responding to your call will connect you with a volunteer doctor at the emergency department of a local hospital (that is local to the agency handling the call, not local to your boat). The doctor will ask you to describe the symptoms of the casualty and the medical equipment and supplies you are carrying onboard. There have been cases where solo sailors have carried out minor surgical procedures on themselves under the guidance of these doctors.

paraffin

a liquid fuel used for lamps, stoves, and heaters; see STOVES (COOKERS).

parallel rules

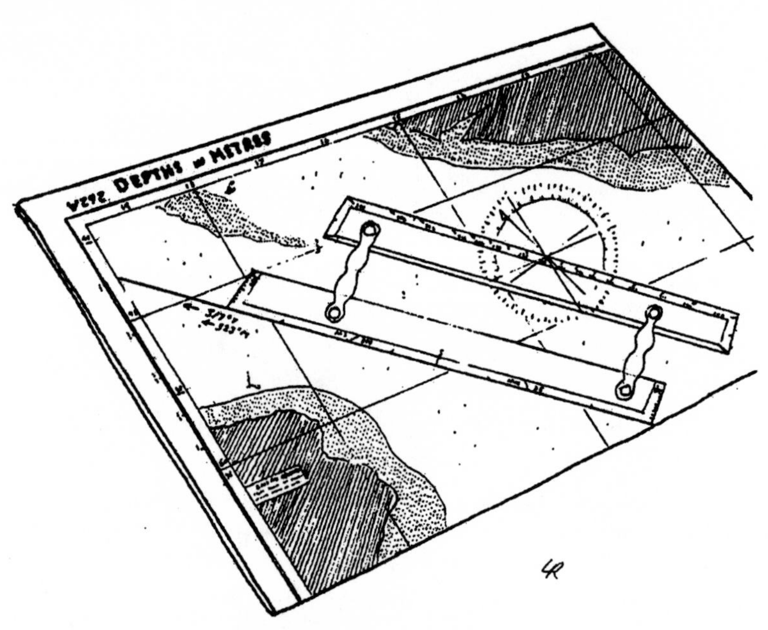

a traditional navigator's instrument; used to draw a COURSE from a COMPASS ROSE on a CHART and then "walk" it across the chart by opening and closing the rules until the start point of the course has been reached. Alternatively, a course can be drawn from the latest fix to the desired next position (off a headland for example) and then "walked" to the rose to determine the true course to be followed. Normally all this is done using degrees true (°T) but some navigators prefer to work in degrees magnetic (°M). The disadvantage of the latter approach, in my view, is that you have to think about the local VARIATION all of the time instead of just some of the time.

parcel, to

to bind a rope with tape or strips of canvas as a means of keeping the water off it.

passage-making

in the case of crossing oceans, a fascinating combination of claustrophobia and agoraphobia:

> In those vast solitudes in the Pacific the feeling is often overwhelming to any thinking man. I have been for four months without seeing even a sail. Nothing but the fish and the waters, and often very few of the fish. The common seamen are often very oppressed by the long, long solitudes.
>
> Anonymous ship's carpenter, 1850

patent log

see LOG.

pawl

another name for a ratchet; a device that enables something (such as a WINCH) to move in one direction but not the other.

pay off

to ease away to LEEWARD while TACKING; compensates for the loss of speed caused by turning through the wind.

pay out

1. to slacken a line so that it can run out freely.
2. as in boatyard, marina office, or chandlers; the least enjoyable aspect of boat ownership.

pedestal

a column that supports a ship's WHEEL.

pennant, a

a tapering flag on a ship, especially one flown at the masthead of a vessel in commission.

perfect boat

an unattainable dream.

The solution to this challenge is, unsurprisingly, dependent upon your financial means. American banker John Pierpont Morgan observed that "Any man who has to ask about the annual upkeep of a yacht can't afford one." So, if money is *not* an issue, do the following:

- scribble your basic requirements concerning performance, accommodation, number of crew, size of helicopter pad, on the back of an envelope;
- write a large check and place in inside the envelope;
- take the envelope to a yacht designer and ask him to come up with some sketches over the weekend;
- on your return from Paris, view the sketches, make some halfway intelligent observations, and have him redo the sketches;
- repeat the previous step a few times;
- once happy, commission a builder and write another even larger check; and
- while the yacht is being built, have a New York–based interior designer start work on the accommodation.

Jack London ran foul of the "money is no object" philosophy when building the *Snark*.

> "Spare no money," I said to Roscoe. "Let everything on the *Snark* be of the best. And never mind decoration. Plain pine boards is good enough finishing for me. But put the money into the construction. Let the *Snark* be as staunch and strong as any boat afloat. Never mind what it costs to make her staunch and strong; you see that she is made staunch and strong, and I'll go on writing and earning the money to pay for it."
>
> And I did . . . as well as I could; for the *Snark* ate up money faster than I could earn it. In fact, every little while I had to borrow money with which to supplement my earnings. Now I borrowed one thousand dollars, now I borrowed two thousand dollars, and now I borrowed five thousand dollars. And all the time I went on working every day and sinking the earnings in the venture. I worked Sundays as well, and I took no holidays. But it was worth it. Every time I thought of the *Snark* I knew she was worth it.
>
> *The Cruise of the Snark* by Jack London

If money is an issue after all, try the following steps:

- take advice from anyone who will offer it;
- read a lot of books and magazines;
- think hard about what kind of sailing you are going to do (coastal, blue water, single-handed, short-handed, family, including the in-laws, and the German Shepherd);

- decide how big the hull will need to be and what it should be made of;
- buy a second-hand boat for the hull (or hull-and-deck) only and rip out the accommodation, wiring, plumbing, and rigging;
- now you can get at the engine, have it professionally checked out and, if it is in good-enough shape, have it removed, stripped down, and restored to pristine condition;
- have a yacht designer take your ideas for the accommodation and convert them to working drawings for the hull;
- with the designer, break the work down into jobs that you can do yourself and jobs that you cannot do yourself (or should be done anyway by a craftsman); and
- where someone else with better skills than yours is doing a job, work for him or her as a hod-carrier so you can get to learn what is involved (this will help should you need to make repairs at sea).

Of course, the more you do yourself, the cheaper it will be and the more likely you will be able to carry out your own maintenance and repairs. There are variations on this theme. For example, a lot of people start to build their own yachts in their backyards and then quit—either because they underestimated the task or money is needed for the divorce settlement. When buying second hand—especially if the hull is home-built—get it surveyed before committing. If you have the extra money, you can buy a yard-built hull and DECK and start from there.

You will have noted that I have not recommended buying an "off-the-shelf" (or "stock") boat because all my calculations show these to be the least good value for money, especially when one considers that a second-hand tub often comes with lots of "extras" such as a KEDGE, mooring lines, fridge, and so on. A new boat has the minimum of these that the builder can get away with. Nor am I enthused by the idea of completely building a yacht from scratch. I don't trust my own skills to that extent and, in any case, I want to go sailing. The answer lies in a compromise between the two; but then I'm English.

personal flotation device (PFD)

See LIFE JACKET.

personal watercraft (PWC)

a small, lightweight planing "craft" designed to be either sat on or stood on with handlebars and a twist throttle, usually jet-propelled; the motorcycle of the wet-ways and potentially just as annoying for grumpy old people like me.

person in water (PIW)

the politically correct version of "man overboard." But a politically incorrect skipper once suggested that it really meant "Pushed In Wife." He was only kidding of course. . . . However, the buttons on GPS sets still say "MOB," so that term will probably be around for a while.

petrel

A particularly fascinating sea bird named, supposedly named after St. Peter:

> And when Peter was come down out of the ship, he walked on the water to go to Jesus.
>
> Matthew 14:29

The petrel flies very close to the water, sometimes kind of paddling on it with its feet. Get the allusion? Charles Darwin liked the look of them, too:

> These southern seas [the Pacific off Chile] are frequented by several species of Petrels: the largest kind, *Procellaria gigantea*, or nelly (*quebrantahuesos*, or break-bones, of the Spaniards), is a common bird, both in the inland channels and on the open sea. In its habits and manner of flight, there is a very close resemblance with the albatross; and as with the albatross, a person may watch it for hours together without seeing on what it feeds. The "break-bones" is, however, a rapacious bird, for it was observed by some of the officers at Port St. Antonio chasing a diver, which tried to escape by diving and flying, but was continually struck down, and at last killed by a blow on its head. At Port St. Julian these great petrels were seen killing and devouring young gulls. A second species (*Puffinus cinereus*), which is common to Europe, Cape Horn, and the coast of Peru, is of much smaller size than the *P. gigantea*, but, like it, of a dirty black colour. It generally frequents the inland sounds in very large flocks: I do not think I ever saw so many birds of any other sort together, as I once saw of these behind the island of Chiloe. Hundreds of thousands flew in an irregular line for several hours in one direction. When part of the flock settled on the water the surface was blackened, and a noise proceeded from them as of human beings talking in the distance.
>
> There are several other species of petrels, but I will only mention one other kind, the *Pelacanoides Berardi* which offers an example of those extraordinary cases, of a bird evidently belonging to one well-marked family, yet both in its habits and structure allied to a very distinct tribe. This bird never leaves the quiet inland sounds. When

> disturbed it dives to a distance, and on coming to the surface, with the same movement takes flight. After flying by a rapid movement of its short wings for a space in a straight line, it drops, as if struck dead, and dives again. The form of its beak and nostrils, length of foot, and even the colouring of its plumage, show that this bird is a petrel: on the other hand, its short wings and consequent little power of flight, its form of body and shape of tail, the absence of a hind toe to its foot, its habit of diving, and its choice of situation, make it at first doubtful whether its relationship is not equally close with the auks. It would undoubtedly be mistaken for an auk, when seen from a distance, either on the wing, or when diving and quietly swimming about the retired channels of Tierra del Fuego.
>
> *The Voyage of the Beagle* by Charles Darwin, 1845

See also MOTHER CAREY'S CHICKENS.

petroleum jelly (Vaseline®)

usually known by its trademark name, Vaseline, this is very useful stuff to have onboard. Here are just a few of its applications:

- spread liberally on your battery terminals it helps prevent corrosion;
- spread less liberally on your feet before getting your socks and boots on, Vaseline helps to prevent "trench foot," a nasty condition caused by having wet feet for too long;
- spread thinly on your fresh eggs before repacking and stowing them increases their shelf life considerably;
- when applied to the end of a rubber or plastic tube being used to administer a freshwater enema (in a survival situation), will be appreciated by the recipient (see WATER, WATER);
- also helps soothe chapped hands.

PFD

see PERSONAL FLOTATION DEVICE

phonetic alphabet

internationally agreed alphabet for use in all radio communications; familiarity with the alphabet is a requirement the UK Radio Communications Agency's Maritime VHF Certificate of Competence. (In the US all you need to know is how to stick a postage stamp on an envelope containing the FCC application form.)

alpha	golf	mike	sierra	yankee
bravo	hotel	november	tango	zulu
charlie	india	oscar	uniform	
delta	juliett	papa	victor	
echo	kilo	quebec	whiskey	
foxtrot	lima	romeo	x-ray	

Do not say "B-bravo," just say "bravo." There are also phonetic phrases for numbers, but these are so bizarre ("uno-wun," "due-too") that I have never heard them used.

piling

heavy post driven into the ground below the WATERLINE to support a pier, or jetty; often something positioned so you can secure a line to it.

pilot

a coastal navigator qualified to guide vessels into and out of ports and harbors and through other restricted waterways such as canals. The word comes from the Greek *pedotus*, rudder, HELMSMAN, through the Latin *pilotus*.

The more incomplete are charts and navigation aids, the more important is the pilot. The audacious Sir Francis Drake appreciated this when, in 1577, he "acquired" a 40-ton smack from the Portuguese. The smack—a cargo vessel—was laden to the gunwales with sherry and Madeira wine. Also onboard was Nuñez da Silva, a pilot with unparalleled knowledge of the Brazilian coast. For Drake this was the equivalent of discovering a GPS receiver and da Silva was "invited" along for the rest of the voyage. So was the sherry and Madeira.

In 1886 Captain Joshua Slocum was less fortunate in his choice of a pilot to guide him through the shoals of Martin Garcia Bar in the River Plate:

> A small schooner captain, observing that we needlessly followed in his track, and being anything but a sailor in principle, wantonly meditated mischief to us. While I was confidently trusting to my pilot, and he (the pilot) trusting to the schooner, one that could go over banks where we would strike, what did the scamp do but shave close to a dangerous spot, my pilot following faithfully in his wake. Then, jumping upon the taffrail of his craft, as we came abreast the shoal, he yelled, like a Comanche, to my pilot to: "Port the helm!" and what does my mutton-headed jackass do but port hard over! The bark, of course, brought up immediately on the ground, as the other had planned, seeing which his whole pirate crew—they could

> have been little less than pirates—joined in roars of laughter, but sailed on, doing us no other harm.
>
> By our utmost exertions the bark was gotten off, not a moment too soon, however, for by the time we kedged her into deep water a *pampeiro* was upon us. She rode out the gale safe at anchor, thanks to an active crew. Our water tanks and casks were then refilled, having been emptied to lighten the bark from her perilous position.
>
> Next evening the storm went down, and by mutual consent our mud-pilot left, taking passage in a passing river-craft, with his pay and our best advice, which was to ship in a dredging-machine, where his capabilities would be appreciated.
>
> *Voyage of the "Liberdade"* by Capt. Joshua Slocum, 1894

See also SAILING DIRECTIONS.

pilotage

coastal navigation using visual contact with sea and land marks; another fine art.

Tip: Become an ace pilot! Improvise a chartboard to use on DECK. Use plywood cut to half-chart size (quarter-chart on smaller boats) and covered with a clear plastic sheet secured with elastic straps. Once you have drawn up your course (before departure) slide the chart under the plastic sheet. Use a chinagraph pencil to add PILOTAGE notes to the sheet. Then position the board on deck; on the cabin roof beneath the spray-hood is a good place. I have seen this arrangement done with an upper WASHBOARD, which is a practical idea, but on some boats the lower board is a bit too small for the job.

pilot house

an enclosed HELM compartment; usually a structure built for the purpose on deck. Sometimes called a wheelhouse.

pinching

pinching (or stealing) wind takes place when a vessel is trying to sail closer to the wind than might be advisable (and with a commensurate sacrifice of speed); a light shift in wind direction could result in the yacht falling IN IRONS. Sailing further off the wind is, BY AND LARGE, to be preferred.

pintle

a means of attaching a RUDDER to the STERN of a boat in such a way that it is hinged and able to move laterally; in simple terms two or more vertical pins on the RUDDER engage in metal rings fitted to the TRANSOM.

piracy, pirates, and privateers

"piracy" is an unlawful attack against a ship on the high seas for the purposes of robbery or kidnapping; a form of mugging at sea. Traditionally performed by leaping across the bulwarks with flintlock pistol in free hand and cutlass between shiny teeth.

Hollywood has always tried to depict pirates as the Robin Hoods of the sea. (This enabled the studios to save money by using the same men's tights in both productions.) However, many "pirates" were actually "privateers," armed sailing vessels legally commissioned by governments to attack and plunder enemy ships in return for a "slice of the action." (See CORSAIRS for more on this.) One of the most famous pirates, the Scotsman William Kidd (c. 1645–1701) began his career as a privateer for the British against French ships off the North American coast. Ironically, in 1695, Kidd was granted a royal commission to put down piracy in the Indian Ocean. But the swashbuckling gamekeeper turned poacher again when he signed up with gang of pirates in Madagascar off the east coast of Africa. He was apprehended in the North Atlantic in 1699, taken to London, and hanged at Executive Dock on the River Thames.

Privateering was made illegal by the 1856 treaty known as the Declaration of Paris. Piracy continues today, especially in the Far East where some coastal communities of Indonesia and the Philippines live off the spoils of armed plunder in the Straits of Malacca and the South China Sea. These pirates swing over your side with RPG-7s and AK-47s between their teeth. In Hong Kong in 1995, I followed reports of piracy by ships of the Chinese Navy moonlighting as privateers. Other risk areas are the Caribbean, the Horn of Africa, and the northerly coasts of South America.

On a global scale, the IMO (via its Piracy Reporting Center in Kuala Lumpur) recorded nearly 1500 incidents in the 1990s. In Southeast Asia there was an increase from 90 to 262 reports between 1994 and 1997 alone. The targets tend to be merchant vessels, whose cargo and their crews' documents can be eminently sellable. The cargo can often be disposed of for anything up to $2M. Such are the remarkably slack regulations concerning the registration of merchant vessels that the ships themselves can also be sold; with a set of sanitized documents, they go for anywhere between $5M and $10M.

In mid-2013 piracy was reported to be in decline around the Horn of Africa, a development that probably has much to with the decision of ship owners to deploy armed force on vessels carrying high-value cargos. In the context of this kind of business, humble cruising yachts are mostly considered to be unworthy of the modern pirates' bold endeavors.

Sadly, there is no real historical evidence that pirates ever flew the "skull-and-crossbones" ensign—that is probably an invention of Errol Flynn's production designer. See also BARBARY COAST, BUCCANEERS, CORSAIRS, and FIREARMS.

pitch

1. the FORE-AND-AFT "nodding" movement of a vessel.
2. the distance, in theory at least, that a PROPELLER would travel in one revolution.

plane sailing (plain sailing)

sailing using navigation techniques that assume the earth is flat. Over long distances (say 500nm or more) it may be necessary to assume that the earth really is round and follow a GREAT CIRCLE course. This involves the use of spherical trigonometry. Often misspelled as "plain sailing" meaning something easy, simple, or basic.

planing hull

a hull designed to ride on top of the water rather than through it; see DISPLACEMENT HULL.

plot

to determine a COURSE using a navigation CHART and a position marked on the chart with a pencil.

point

a point is 12.25°; see COMPASS POINTS.

pooped

having a great big wave come over the STERN; always happens when (a) you least expect it (b) when the WASHBOARDS are out and (c) the CHICKEN SUPREME is ready. In everyday English means "very tired."

port

the left-hand side of a vessel when facing forward; used to be called the "larboard" side because it was where the LEEBOARD was fitted. In pre-RUDDER days the right-hand side, "starboard" was where the steering-board was and, therefore, could not be jammed against the pilings.

port light

a small opening window set into the TOPSIDES, COACH-ROOF, or COAMINGS of a boat; sometimes called a "port hole."

port tack

sailing CLOSE-HAULED with the wind coming over the PORT side of the boat.

posh

"port-out, starboard-home"; believed to date from the days of the British Empire which ships en route to India charged extra for port-side cabins, which were away from the heat of the sun. On the return trip it was the starboard-side cabins that attracted a premium. This is a nice story but, according to historians of the era, it is completely untrue.

position line

a line drawn on a chart by a navigator (may be derived from a BEARING); the boat may be at any point along a position line. Where two position lines intersect, the boat's position may be fixed at that intersection. A third line provides reassurance, even if a COCKED HAT is formed.

powder monkey

on square-rigged men o' war boys were employed to carry black powder from the ship's magazine to the gun DECK; generally, the smallest of anything on such a vessel was called the "monkey," hence "powder monkeys."

pram

a small double-ended DINGHY with a flat TRANSOM and BOW.

predictions

"We are as near to heaven by sea as by land," said Sir Humphrey Gilbert. And then the *Squirrel*, his ship, sank and Sir Humphrey drowned. An earlier prediction—"I'm going to Newfoundland and claim it for Britain!"—turned out to be true, too. Two out of two is not bad going. Sir Humphrey (c. 1537–1583) was half-brother of Sir Walter Raleigh.

preventer (boom preventer)

a line that connects a BOOM to the HULL on the opposite side of the sheet; used to prevent accidental GYBING, especially when RUNNING downwind in heavy seas. Also used to describe any device that has the purpose of stopping something dangerous happening.

privateer

privateers were privately owned ships commissioned by their governments to attack and seize enemy cargo vessels. By comparison PIRATES did not bear anyone's warrant. Privateers were the forerunners of today's private military companies.

propane gas

see STOVES (COOKERS).

propeller

a means of transferring the power of a ship's engines (or a yacht's auxiliary engine) to forward propulsion through the use of (two or three) angled blades fitted to the end of a shaft emerging from the STERN of the vessel.

Like most people, I had two grandfathers. One was a Master Mariner who decided to jump ship in Alaska and open a hotel. I never knew him. My other grandfather was a Forge Master who, in the 1930s, worked in the shipbuilding industry. At the beginning of World War II, my father, like most other young men, found himself in uniform. He became a driver in an artillery regiment in the 8th Army and, in 1940, was dispatched to the port of Glasgow in Scotland. On the pier he and his comrades looked up at a huge passenger liner—it was the *Queen Mary*. She had been repainted in navy grey, but there was no mistaking her three massive funnels. Thousands of troops were packed below decks before she steamed out of the Clyde and headed south down the Irish Sea toward the North Atlantic and U-boat territory.

Off the coast of West Africa, the liner began to outstretch her destroyer escort and she was left to continue her journey south alone. When she arrived unscathed at the Simonstown naval base in South Africa, speculation was rife among her passengers that they were headed for India. But on leaving the Cape the *Queen Mary* (and my father) headed north toward the Mozambique Channel. Intelligence reports had placed wolf packs of German submarines in the area and, on reaching the Channel, the *Queen Mary* picked up her skirts and made all speed.

She was able to achieve an astounding forty knots, but not without some discomfort to the passengers. Severe vibration made conditions on the crowded lower decks even worse. After a day of this hot, humid, and noisy torment, the soldiers began to fear the ship more than the marauding German U-boats.

"What if the bloody propellers drop off?" asked one pessimist.

"They won't," said my father from his bunk.

"How the hell do you know?"

"Because my dad made them."

The ship eventually made it through the gauntlet and fetched up at Suez. My father was given a truck and a Thompson submachine gun and set off to play his part in Rommel's downfall in the deserts of North Africa.

I have also been aboard the *Queen Mary*—in Long Beach, California, where she is permanently installed as a hotel and conference center and limited to zero knots. I do not know if my grandfather's propellers are still attached.

prop walk

a boat's propeller, as well as making the hull move forward or backward, also has a tendency to make it move to one side; this is especially true when in reverse. This characteristic is both a nuisance and benefit. For example, when mooring alongside, prop walk can be used to kick the STERN in toward the berth. Similarly, when leaving a berth, prop walk can be used to move the stern out toward the open water.

proverbs

whoever wrote this bit of the Bible was certainly missing out:

> There be three things which are too wonderful for me, yea, four which I know not: The way of an eagle in the air; the way of a serpent upon a rock; the way of a ship in the midst of the sea; and the way of a man with a maid.
>
> Proverbs 30:18–19

But things move on and even I can claim to "know" four out of these four wonderful things. . . . See also EYES.

provisioning

the acquisition by purchase or other means of food, water, and other consumables for an imminent passage over the sea.

In the days of sail, with large crews and little certainty about when they would next make land, the matter of provisioning was paramount. This is a story told by Arthur Knights about a voyage undertaken in the winter of 1854–1855.

Well, I joined the barque "Tropic," loaded with guano, bound for Cork, in Ireland. This vessel was a very rotten old thing, and in getting round Cape Horn we all had a very hard time, and did not know how soon the vessel would sink with us; but we got round the Cape and into the South Atlantic, where we had better weather and proceeded pretty well till in the North Atlantic, when provisions began to get short. When we were off the Azores, watching the beautiful shores and harbours of St. Michael, we came near a Dutch brig from Brazil loaded with coffee. The captain hailed us and asked us for some biscuits. A boat was sent to us bringing us a half-bag of coffee. We had less than a hundred pounds of biscuits. Our captain consulted with us about giving any of it away. It was finally agreed that we would divide with the brig. This was done, and we had to be very careful with so little bread among twelve people. We had plenty of salt beef and pork, and a half-barrel of flour, but no beans or peas or sugar.

We had a fair run till we saw Cape Clear, at the south end of Ireland, on the 30th of January, 1855. We all were in high hopes that a few hours more would see us at anchor in Queenstown; but that night came on an easterly gale, and we were driven out into the Atlantic, where for weeks we were buffeted about, and to our dismay our last fresh-water cask we found had leaked and was empty. We were surrounded with many other vessels in the same plight—short of provisions. We had plenty of snow, with which we could make coffee, but were reduced to salt meat only, which is pretty hard fare. The hardest part was, that the captain had his wife and two children on board, and for the youngest child a goat had been provided to supply milk. This became a scarce article as there was no food for the goat. So every day the carpenter used to plane up a piece of wood to make shavings for the goat to eat. It got along as well or better than any of us.

Notes by the Way in A Sailor's Life
by Captain Arthur E. Knights, 1898

See HARD TACK and CHICKEN SUPREME.

prow

the forward part of a boat above the WATERLINE.

pulley

a wheel with a grooved rim around which a LINE can be turned; from the Old French word *polie* and first seen used in a nautical context in 1295.

See also BLOCK AND TACKLE.

pulpit

a guard rail fitted to the BOW of a yacht; analogous to a pulpit in a church; see PUSHPIT.

pump

a mechanical device working on the suction principle that can move water from where it is now to where you want it to be (e.g., water from the BILGES and into the sea or water from the sea into the HEADS). Can be hand driven or electrically powered. Total reliance on electrical pumps is unwise; the reason that you need to pump might also be the reason that the pump doesn't work anymore.

pump well

a depression in the lowest part of the BILGES where water can gather and be pumped overboard; sometimes called the "sump."

punch

a brew of rum or brandy and wine served hot from a bowl or other container; may be the original maritime "cocktail."

Any old booze will do when the Royal Navy is in port. Punch is a concoction that originates with the ANDREW (as the navy is known in British military circles) and supposedly dates back to the mid-1500s. Now forget about the kind of thing a bartender is likely to serve you in a glass full of ice; the original comes hot and very alcoholic in a very large bowl. Admiral Sir Edward Kennel set the standard:

> On the 25th October, 1599, Sir Edward Kennel, commander-in-chief of the English navy, offered to his ships' companies a monster punch which he had prepared in a vast marble basin. For this concoction he used 80 casks of brandy, 9 of water, 25,000 large limes, 80 pints of lemon juice, 13 quintals (1,300 pounds) of Lisbon sugar, 5 pounds of nutmeg and 300 biscuits, plus a great cask of Malaga [wine].
>
> A platform had been constructed over the basin to shelter it from the rain, and the serving was done by a ship's boy who sailed on this sea of punch in a rosewood boat. To serve the 6,000 guests one ship's boy had to be replaced by another several times, each one finding himself intoxicated by the fumes from the lake of alcohol at the end of a quarter of an hour.
>
> *Larousse Gastronomique*

purchase

another name for a TACKLE.

pushpit (UK)

a guardrail fitted to the STERN of a yacht above the TRANSOM; assumed to be a joke: "pulpit" at the front, "pushpit" at the STERN (in America pushpits seem to be referred to as "stern pulpits" which is not nearly as funny); see PULPIT.

PWC

see PERSONAL WATERCRAFT.

Q

Q flag

yellow signal flag for the letter Q "quebec" and having the secondary meaning: "My vessel is healthy and I request free *pratique* [clearance certificate]." It is flown at a port of entry to request customs, immigration, as well as health authority clearance.

The usual procedure is this: raise the Q flag to the starboard CROSSTREES (SPREADERS) on entering the country's territorial waters. As soon as possible go ashore and get clearance. In some places Customs and Immigration will visit you at ANCHOR or at your marina berth. You must not go ashore until you have cleared in—or you are going ashore in order to clear in. This process usually involves filling in a lot of boring forms for the boat and for each member of the crew. In some places the skipper can take the ship's papers and the crew's passports and do this on his own. In other places, you need to go mob-handed. Either way, take money; there will be cruising and fishing permits to buy and fees for the visas. They will want to know if you have FIREARMS onboard and they might take them into safe-keeping for the duration of your visit. Customs may want to search the boat, especially if they do not like the look of you. In some areas this will be done aggressively; they will damage the boat and laugh when you ask about compensation. Once you have cleared, you take the Q flag down and replace it with the COURTESY FLAG.

These procedures vary from country to country, so read the relevant PILOT or sailing directions and get it right; border officials have a low boredom threshold and can be very irritating when they feel so inclined. In some countries you are supposed to fly both Q flag and courtesy flag on entry, then just take the Q down on clearance. If you have to hand in the guns, you will need to formally leave the country via your port of entry so you can get them back again. You can only enter at an approved port of entry so do not go EXPLORING or GUNK-HOLING with just the Q flag up or you will be in trouble. In Greece, though, you will be met by a customs official who, as you tie up, will tell you to go to his office, not entering any bars that you might pass on the way. This happens at *every* island.

I hear that the European Union has simplified matters but I have not experienced the process yet. I only hope the French have been told about the new rules.

quarter

generally, the aft port and starboard corners of a vessel. Anything abaft the beam is referred to as being on the port or starboard quarter.

quarter berth

a berth built into either quarter of a sailing boat. Usually shaped like a coffin (and rarely much bigger) and entered via one end (inside the vessel, of course). Can be snug. Can also be oppressively hot in tropical climes (see VENTILATION). A window opening through the hull or to the cockpit is recommended, so long as it can be guaranteed to be watertight. Quarter-berths can often be drenched through open COMPANIONWAYS unless care is taken. See BIVOUAC BAG.

quartering

the tactic, when working to windward in heavy weather, of steering the BOW at 45 degrees to the oncoming waves.

quarters

the living and sleeping accommodation of a boat or ship.

quay

a wharf, traditionally used for the loading and discharge of cargo.

R

rabbet (UK: rebate)

a slot or groove cut into one piece of timber in order to receive the edge or end of another.

racer

a sailboat designed for speed rather than comfort or, some might say, safety; see TOOTHBRUSH, HALF A.

racing

the sport of matching the performance of similar boats over a predetermined course; can take place in DINGHIES or ocean yachts over 20m LOA.

This was once a serious business when PILOTS would race each other to be first to reach arriving vessels that would need their services before being allowed to enter port. Still a serious business with heavy sponsorship or fee-paying crews competing in global circumnavigations. Ocean racing is like standing in a cold shower tearing up £50 notes. It might possibly be enjoyable if they were someone else's £50 notes, but I doubt it.

I was once a very keen, albeit over-aggressive, dinghy racer. This all came to an end when I realized that the chaps beating me in Fireball dinghies each week were out jogging every morning to get fit. When I spotted one of them doing sit-ups off the side of a bench to strengthen his belly muscles, I there and then wrote out a for-sale notice and pinned it to the club notice board. See LOCAL KNOWLEDGE.

Racundra

Racundra was a 30ft center-board ketch commissioned by Arthur Ransome for cruising in the Baltic Sea. (Ransome is of course famous worldwide for his children's books, especially *Swallows and Amazons*.) In 1923 Ransome published a factual account of the boat's maiden voyage, *Racundra's First Cruise*. In the book he says that his crew comprised a female cook and someone he called the "Ancient Mariner." He claimed to have first met the Ancient Mariner when the latter was a harbor master "of Baltic origin" in the Latvian port of Riga. This is what he wrote about him.

> Many, many years ago he sailed on the famous *Sunbeam* of Lord Brassey. He had been a seaman on the *Thermopylæ*, which he called the "Deemooply," and had raced in her against the *Kutuzak*, in which odd Russianised name I recognised the *Cutty Sark*. . . . "I am an old man," he said, "and I should like once more to go to sea before it is too late." And I, of course, agreed with joy, for there is no such rigger in the Baltic as the Ancient Mariner who has known what it was to sail in the *Thermopylæ* in the days of her pride.

There is no reason to believe that most of Arthur Ransome's account of the first cruise of the *Racundra* is not entirely credible. However, you need to be aware that, for some years now, there have been suggestions that this former foreign correspondent of the Manchester *Guardian* newspaper had also been operating as an MI6 (Secret Intelligence Service) agent spying against the new Bolshevik government in Russia. Ransome did indeed claim to be a close friend of Lenin and Trotsky and he later married Trotsky's secretary. Papers recently discovered confirm these allegations and even appear to solve the mystery of the Ancient Mariner. The third crew member onboard *Racundra* was none other than Ernest Boyce, Ransome's controller in MI6. Not a very Baltic name, Boyce.

radar

an electronic device using giga-high frequency radio waves to detect objects and display their "echoes" on a CRT or LCD monitor; new digital radar systems have become increasingly affordable and have impressive functionality. Digital radar can interface with NMEA networks, making it possible for them to display GPS and even digital chart information over the radar picture.

radiation fog

land fog caused by the condensation of water vapor above cooler ground; can drift over the sea in coastal areas. See CONVECTION FOG.

radio beacon (or radio station)

a navigation aid that transmits a radio signal (usually medium wave—amplitude modulation [AM]). The signal is arranged in such a way that the beacon can be uniquely identified, by sending a group of identifying letters in Morse Code. A receiving set incorporates a compass so that a BEARING can be taken. The above two sets of information can then be combined to enable a POSITION LINE to be drawn on a chart. The vessel of the observer will then be somewhere on that line. This process is known as "radio direction finding" or RDF. By adding at least one additional bearing line that intersects with the first, it is possible to obtain a FIX. The additional bearings may be obtained by visual means or by taking another RDF using a radio beacon at a great enough angle to the first from the vessel's approximate position. Some radio beacons have been set up for the purpose of maritime navigation (see LANBY) and these are supplemented by beacons at airports. All very low-tech compared with GPS, but useful as a double-check.

raft

a rudimentary type of vessel fashioned from floating materials such as bamboo or balsa wood lashed together and propelled by ROWING, SCULLING, or SAILING. I have always been fascinated more than intrigued by adventurers such as Thor Heyerdahl and Tim Severin who, through their determination to prove that ancient peoples routinely traveled from one continent to the other to pick up the groceries, more often proved that the raft was a dead loss when it came to serious cruising. The tension in the accounts of their voyages comes not from the question: Will we get there? but from the question: How far will we get before this useless thing falls apart and we push the button on the EPIRB?

Perhaps the most charming such adventure was undertaken single-handedly in 1954, seven years after Heyerdahl's *Kon-Tiki* Expedition. This was the largely forgotten *Epic Voyage of the Seven Little Sisters* in which William Willis, an American merchant seaman, built and sailed a balsa raft 6,700 miles from Peru to Pago Pago. *Kon-Tiki* was crewed and only covered 4,300 miles. Willis was accompanied by a black kitten called Meekie and a parrot called Eekie. The cat caused problems the whole trip and Willis put his life at risk on a number of occasions to stop it going overboard. Seeing it was so keen to become shark bait, I would have had it at the end of a line before the Andes had dipped below the horizon.

> Today, June 30, I had been out a week and saw blue sky for the first time, a little patch no bigger than a fingernail. That meant I had not been able to take a sight and so could only guess at my position. By dead reckoning I put myself at Lat. 8°17' South and Long. 82°20' West, 400 miles from Callao. But there was no question about being in the Humboldt Current. It was cold. I had to wear two sweaters, a flannel shirt and a pea-jacket to keep warm, also two pairs of woollen socks Teddy [his wife] knitted for me. It had been cold since the day I left Callao towed by the *San Martin*.
>
> How strangely everything drifted out of my mind, as if I had no real contact with the world of men, no real hold on its existence. I could hardly remember anything and details not at all. It disturbed me. My sense of time had disappeared also.
>
> As the day waned I filled my lantern with kerosene to light up the compass and got ready for the night. The old oilskin coat went around Eekie's cage to keep out the penetrating cold. Meekie was all snug underneath the winch on an old sweater. After a short twilight it was dark, as is the way of the tropics.
>
> *The Epic Voyage of the Seven Little Sisters*
> by William Willis, 1956

You just knew the damned cat was going to get the upper hand. Near the end of the voyage, Willis discovers the moggy licking its lips next to an empty cage and a pile of bright green feathers.

As I write this particular entry (August 1998), the latest epic transocean RAFT voyage has just ended in a successful landfall at Castletown, Co. Cork, Ireland. The oddly named *Son of Town Hall* has to be the strangest-looking vessel ever to touch water. The last time I saw anything remotely resembling it was in Kevin Costner's movie *Waterworld*. *Son of Town Hall* is actually—if you can imagine this—a boat-shaped raft rigged as a square-sail junk-rigged schooner. The rest of the 50ft vessel seems to have been rigged from junk too. "It looks like two sails on top of a garden shed," was the somewhat-flattering comment from one observer. The local harbor-master was reported as saying that "It looks like something out of the *Beverly Hillbillies*." I bet it was unimpressive to windward. It had, however, been blessed as seaworthy by the US Coast Guard and had an auxiliary engine, two VHF sets, two short-wave transceivers, RADAR, and the usual flares and PFDs.

Needless to say, it also had an interesting crew. According to a report in the London *Guardian* they were "Four artists too poor to fly, but keen to see the world." They had "crossed the Atlantic (from Nova Scotia) with three dogs aboard a boat made of recycled plywood and barrels." The crew comprised Poppino and Aurelia Neutrino, a couple in their sixties from San Francisco; a Canadian called Roger Doncaster; an Edward "Mr. Garry" Garry; two Rottweilers, Siegfried and Thor; and a Mexican poodle called Willy. They all lived off pasta and tinned stuff but were short of food and water on arrival. The planned trip of thirty days had taken sixty-three.

These wonderful characters were planning to move on to the Mediterranean and I wish them Godspeed.

Tip: If you want to build a boat from recycled materials, try recycling another boat.

rafting

a system of mooring using limited berthing space, based on tying up to other boats. Good manners dictate that such requests are readily acceded to but often lead to an instant decision to leave on the 4.20 am high tide. Sometimes a refusal is the smart decision, though. In the mid-1990s we were making a delivery of a 42ft charter yacht from a boatyard at Lavrion on the Greek mainland to Rhodes. During an easterly trek along the north side of the island the weather started to blow up, so we decided to duck into the tiny harbor at Palon. Mooring there is stern-to against a mole protecting the harbor from the north.

As the afternoon drew on, more yachts followed us in and the space for eight or nine boats was soon filled. Next to us was a yacht crewed by a group of cheerful young Germans whom we'd met the previous day in bar on Astipalaia. While the rest of our crew went for a walk with the Germans, I stayed behind do some laundry. The MELTEMI was gaining in strength all the time. I was busy with the last few T-shirts when a shout came from the mole and I stuck my head out.

By now the swell was hitting the mole and spraying all over my smalls. I looked over the bow to where a fellow crew member was pointing and was amazed to see a fleet of about a dozen yachts all flying the ensign of a Finnish bank. One of them had already wedged itself bow-twixt-bow between us and the Germans. Its crew looked completely exhausted, slumped in the cockpit, still in their oilies and life jackets. We did what we could to guide the others into the available space until only two remained.

I asked the Finns on the boat next to us if the skippers still on the other two yachts spoke English and if they would have their VHF radios on. They shrugged, far too past it to care. Then one of the yachts headed toward the gap between us and the Finnish yacht on the other side. Now we were seriously worried about the ability of our shorelines to hold another yacht. (We were carrying a financial penalty for any damage caused in transit, I'll admit that.) We shooed them off. Right or wrong, we were only twenty meters from the point at which we would *all* have gone aground. One boat rafted up—dangerously in our view—to two others.

Eventually, the last one headed out on the roller coaster coming in between the moles to head back to Mandraki. We watched as a teenage boy skipped around the mast, shaking out a reef, wearing neither life jacket nor safety harness. He looked as though he knew what he was doing and we prayed he did. They got out safely, but our concern remained—Mandraki is even less safe in a Meltemi.

The following morning the wind had dropped and we chanced the roller coaster out of the narrow gap ourselves. The Germans sat in the bar and watched, thinking the Brits were as crazy as the Finns. Wrong. We are much crazier than the Finns.

rail

as its name suggests, a rail that runs around the edge of the DECK, above the bulwarks; on yachts usually just a toe rail to prevent crew from slipping under the LIFELINES.

rake

the FORE-AND-AFT incline from the perpendicular toward the STERN of a mast or funnel.

rakish

looking rather racy and streamlined; from RAKE.

range

1. a tidal range is the difference between high water and low water on any particular TIDE. Low ranges are a character of neap tides; high ranges characterize spring tides.
2. to range an anchor cable (US: RODE) is to lay the cable out on DECK before dropping the ANCHOR. Synonymous with to FLAKE.

3. the distance a boat can travel at cruising speed on one full tank of fuel.
4. the distance from the boat to an object.
5. in intra-coastal navigation, a set of two markers which, when lined up, indicate the deepest part of the channel; in the case of the INTRA-COASTAL WATERWAY this is no guarantee that you will not go aground.

ratlines (pron. "ratlins")

ladder-like arrangements of rope used on square-rigged, lug-rigged, and other larger sailing vessels to enable the crew to climb to the upper reaches of a mast (at least to the SPREADERS).

Tip: Do not look down. Even halfway up, the rest of the boat begins to look like something you last saw in a bathtub.

reach, reaching

sailing a course with the wind broadly coming across the beam; anywhere between CLOSE-HAULED and RUNNING. The most comfortable and often fastest point of sailing.

ready about

the last cry from the helmsman before he throws the wheel over to turn the BOW through the wind; time for the crew to ready the SHEETS and duck before the BOOM comes hurtling across the DECK.

reciprocal course

a course set in exactly the opposite direction for the one presently being sailed.

red

color normally associated with PORT.

redundancy

1. the reason you are on the boat; the company down-sized, your job was "shed" but you got a fair payoff.
2. an engineering concept whereby a system has in-built duplication; the Lagan variation states that every artifact aboard an ocean cruising sailboat should, wherever possible, have a secondary purpose. Here are some examples.
   - A BOATHOOK's primary purpose is to fend off or pull in buoys, piers, or other boats. Its secondary purposes might include a harpoon (with a

different head); a WHISKER POLE; a prop for a cockpit canopy (BIMINI); or a weapon for fending off pirates (especially with a sharpened point).
- No one should go to sea without some means of escape from a burning and/or sinking yacht. But why have *two* dinghies when the functionality of yacht's TENDER and yacht's LIFERAFT can be combined into one? See LIFE-TENDER for more on this.
- Keep a body bag in the LIFE-TENDER. Then, if anyone is suffering from exposure, put them in it—it will help retain their body heat. Also saves on repackaging should they expire.
- A bathtub on a yacht is an unparalleled luxury but you can salve your conscience if it is also being used as a washtub (for clothes), the base of a wet-locker, or somewhere to put wire baskets holding fresh vegetables and fruit.
- Unless you are one of those sailors inextricably and inexplicably drawn to the polar regions (makes me shiver just to type that), do you really need a heater? Why not just put one of the cooker (stove) rings on? If you are worried about safety, rig an inverted wire basket over the rig that is burning to reduce the risk of anything falling directly onto the naked flame. (The basket of a deep fryer with the handle cut off serves well.)
- Remember the first handheld GPS set you bought? Then, once you were hooked, you bought a proper one (or even a GPS PLOTTER) with a big display and a repeater in the cockpit? Well, what happened to the handheld job? If you gave it to the local sea cadets, fine. If you didn't, it is now your backup and lives in the GRAB BAG, just in case.
- Shampoo/shower gel that works with salt water is expensive and not readily available in most parts of the world. However . . . cheap and readily available washing-up liquid seems to do the job just as well. I wonder if it is possible to cook with it? Or top up the engine sump? It must be good for cleaning off those annoying scuff marks on the GRP?

I once rigged a BOATHOOK with a loop for catching snakes.

reef

1. to "reef" is to reduce sail in order to compensate for an increase in wind strength. This is traditionally achieved by lowering the MAINSAIL toward the boom and tying it off using reefing lines attached in rows and at intervals to the sail. More modern sailing boats might be fitted with in-mast or in-boom roller reefing that aims at making the process possible from the cockpit or when short-handed. Both methods add cost and complication and, in my experience, rarely work in a foolproof manner. In the case of the FORESAIL, most cruising yachts have a reefing arrangement based on

rolling the LUFF of the sail around the FORESTAY. This largely replaces the madness of having to go onto the FOREDECK and replacing JIBS entirely with a smaller (or larger) one. I wonder how many sailors have lost their lives doing this in severe weather conditions?

2. a reef is a mostly underwater ridge of rock. Usually, the kind of coral structure encountered in the tropics comes to mind, but that needn't be the case. Both uses of the word reef originate from the Old Norse *rif* meaning rib. The word may refer to the horizontal seam of a sail or to the position of a batten. The noun became a verb in the 17th century. In the case of the "underwater ridge of rock" it may be to do with the shape of the reef or with the fact that it is just concealed under the surface of something.

reeve, to

to run a rope through a FAIRLEAD or BLOCK.

reference station (US)

equivalent to a STANDARD PORT in UK tide tables; high and low tides (times and heights) are published for each day for every reference station throughout the year, adjustments being made where necessary for intermediate subordinate stations. See also TIDE TABLES.

restricted vessel

a "restricted vessel" is a "vessel restricted in her ability to maneuver" according to the COLREGS (US: NavRules) International Rule 3 General Definitions:

> g) The term "vessel restricted in her ability to manoeuvre" means a vessel which from the nature of her work is restricted in her ability to manoeuvre as required by these Rules and is therefore unable to keep out of the way of another vessel.
>
> The term "vessels restricted in their ability to manoeuvre" shall include but not be limited to:
>
> (i) a vessel engaged in laying, servicing or picking up a navigation mark, submarine cable or pipeline;
> (ii) a vessel engaged in dredging, surveying or underwater operations;
> (iii) a vessel engaged in replenishment or transferring persons, provisions or cargo while underway;
> (iv) a vessel engaged in the launching or recovery of aircraft;
> (v) a vessel engaged in mine clearance operations;

(vi) a vessel engaged in a towing operation such as severely restricts the towing vessel and her tow in their ability to deviate from their course.

See also CONSTRAINED VESSEL—which is not the same thing.

reverse osmosis

a system of removing salt from water by forcing the water under pressure through a membrane.

A desalination plant makes it possible to produce drinking water as the sea miles are stacked up. This is a reverse-osmosis system with a filter and can produce about 18 liters (just over 5 US gallons) an hour. A pump takes water from the same through-hull source used to cool the engine and forces it into a long tube under great pressure. Inside the tube is a selectively permeable membrane that allows the molecules of water to pass through, but not the salt. There are two outputs; a trickle of drinking water and huge amounts of brine that go back over the side. The drinking water can be fed into the boat's tanks or into empty water containers.

This is a great piece of kit, but expensive and draining on the batteries. It is recommended that it is only used when the engine is running—which is usually long enough, even on a sailing boat. There is also a small hand-operated version of this machine that can be stowed in the GRAB-BAG of survival extras. Having tested this device, I suspect that I was losing more fluid by sweating than I was gaining from the output tube.

RIB

see RIGID INFLATABLE BOAT

riding light

a light or lantern hung at the FOREDECK and showing a white light all round. Required by the COLREGS (US: NAV RULES) to indicate that a vessel is at ANCHOR.

rig

the configuration of a ship in terms of its masts, booms, sails, and lines; in the case of a sailing vessel, it is this arrangement that determines her type (i.e., BERMUDAN, KETCH, SLOOP, LUGGER, etc.).

rigging, the

the arrangement of lines (ropes, cables and chains) used to control the sails of a vessel: the standing rigging is the system of SHROUDS and STAYS used to keep the masts in place; the running rigging is the system of sheets and halyards used to control the sails.

rigging screw

a strong metal fitting with interior threading at both ends; used to secure and tighten rigging cables. See also BOTTLE SCREW.

right of way (power-driven vessels)

Even on the high seas, outside of territorial waters, there are "rules of the road" aimed at the prevention of collisions. These rules have the force of international law (see COLREGS, NavRules). This is how they are formulated.

International Rule 14 Steering and Sailing Rules

Head-On Situation

(a) When two power-driven vessels are meeting on reciprocal or nearly reciprocal courses so as to involve risk of collision each shall alter her course to starboard so that each shall pass on the port side of the other.
(b) Such a situation shall be deemed to exist when a vessel sees the other ahead or nearly ahead and by night she could see the masthead lights of the other in a line or nearly in a line and/or both sidelights and by day she observes the corresponding aspect of the other vessel.
(c) When a vessel is in any doubt as to whether such a situation exists she shall assume that it does exist and act accordingly.

International Rule 15 Steering and Sailing Rules

Crossing Situation

When two power-driven vessels are crossing so as to involve risk of collision, the vessel which has the other on her own starboard side shall keep out of the way and shall, if the circumstances of the case admit, avoid crossing ahead of the other vessel.

International Rule 16 Steering and Sailing Rules

Action by Give-way Vessel

Every vessel which is directed to keep out of the way of another vessel shall, so far as possible, take early and substantial action to keep well clear.

International Rule 17 Steering and Sailing Rules

Action by Stand-on Vessel

a)
  (i) Where one of two vessels is to keep out of the way the other shall keep her course and speed.
  (ii) The latter vessel may however take action to avoid collision by her manoeuvre alone, as soon as it becomes apparent to her that the vessel required to keep out of the way is not taking appropriate action in compliance with these Rules.

(b) When, from any cause, the vessel required to keep her course and speed finds herself so close that collision cannot be avoided by the action of the give-way vessel alone, she shall take such action as will best aid to avoid collision.

(c) A power-driven vessel which takes action in a crossing situation in accordance with subparagraph (a)(ii) of this Rule to avoid collision with another power-driven vessel shall, if the circumstances of the case admit, not alter course to port for a vessel on her own port side.

(d) This Rule does not relieve the give-way vessel of her obligation to keep out of the way.

See also RIGHT OF WAY (SAILING VESSELS).

right of way (sailing vessels)

you are approaching another vessel under sail: who has right of way? All is made clear in ColRegs (US: NavRules) International Rule 12:

a) When two sailing vessels are approaching one another, so as to involve risk of collision, one of them shall keep out of the way of the other as follows:
  (i) when each has the wind on a different side, the vessel which has the wind on the port side shall keep out of the way of the other;
  (ii) when both have the wind on the same side, the vessel which is to windward shall keep out of the way of the vessel which is to leeward;
  (iii) if a vessel with the wind on the port side sees a vessel to windward and cannot determine with certainty whether the other vessel has the wind on the port or on the starboard side, she shall keep out of the way of the other.

b) For the purposes of this Rule the windward side shall be deemed to be the side opposite to that on which the mainsail is carried or, in the case of a square-rigged vessel, the side opposite to that on which the largest fore-and-aft sail is carried.

These rules need to become instinctive. When piloting through busy coastal waters, a skipper should not have to think twice about whether he is the STAND ON VESSEL or the GIVE WAY VESSEL. Having said that, an unthinking application of the NavRules—especially when a small sailboat encounters a super-tanker—could mean that you are not around for the marine accident enquiry to decide if you were right or wrong. The following epitaph says it all:

> Here lies the body of Michael O'Day
> Who died maintaining his right of way;
> He was right, dead right, as he sailed along,
> But he's just as dead as if he'd been wrong.

Think about it.

rigid inflatable boat (RIB)

an INFLATABLE BOAT with a hard bottom; often used in military and search-and-rescue applications.

Roaring Forties

the forty degree latitudes of the Southern Ocean.

> Below the Roaring Forties, there is no Law.
> Below the Fifties, there is no God.
>
> Old proverb

rocket launcher (US)

a nickname for an array of tubes fitted to sport fishing boats to take the handles of fishing rods; also called "rod holders" or the "Stalin's Organ."

rocks, on the

with ice.

rode (US)

the line or chain connecting an ANCHOR to the vessel; probably derives from the expression "to ride to anchor" meaning that the anchor is secure in the bottom and is preventing the vessel from drifting with wind and/or tide.

roller

a wave that is breaking at the top before it hits the shore.

roller-reefing

a popular system of reducing sail by causing the sail to wrap around a rotating BOOM; see also REEFING.

round turn

a round turn is made with a rope when it is wrapped once around a rail to point in its original direction (360°) and then brought back to lie parallel with itself (another 180°); not a knot, but used in knots such as the FISHERMAN'S BEND.

round up, to

to turn head-to-wind from a REACH or RUN.

rowlock (pron. "rollock")

U-shaped slots in the GUNWHALE of a dinghy that serve as a turning point for the oars.

Royal Greenwich Observatory (RGO)

The RGO was established in 1675 by King Charles II. He provided his first Astronomer Royal, John Flamsteed, with the following mission statement:

> To apply himself with the most exact care and diligence to the rectifying of the tables of the motions of the heavens, and the places of the fixed stars, so as to find out the so much desired longitude of places for the perfecting of the art of navigation.

More than three centuries later, Her Majesty's Nautical Almanac Office is no longer based in Greenwich, but remains responsible for the computing, printing, and publication of the NAUTICAL ALMANAC and many other associated services.

rubrail (US), rubbing strake (UK)

protective bumper running around the boat where the top DECK meets the hull or, preferably, at the widest point of the hull; a British sailor would call this a "rubbing strake."

rudder

a long flat(ish) plate hinged to the TRANSOM or KEEL of a vessel and connected to a TILLER, WHEEL, or TIPSTAFF by way of a rudder post. Movement of the tiller or wheel is transferred to the rudder, which then causes the boat to change in direction.

rummage

Customs' word for a search of a ship. Originates from the old word for a ship's HOLD, or cargo storage space.

running

to "run" is to sail downwind (with the wind coming over the STERN of the yacht).

running backstays

BACKSTAYS used to add extra stability to a mast against the persistent motion of a yacht during an extended voyage.

running by the lee

RUNNING downwind with the wind coming from the same side of the vessel as the main boom that can lead to an uncontrolled gybe and is potentially dangerous except in light winds.

running lights

see LIGHTS ON SAILING VESSELS.

running rigging

see RIGGING.

running spring

a SPRING line used to maneuver into or out of a berth.

RYA (Royal Yachting Association: UK)

The "governing body" of recreational sailing in Britain. The RYA says that in Sail Cruising matters it

- caters for the information needs of [its] personal members
- represents the cruising yachtsman in Governmental, non-Governmental and EU matters
- publishes information booklets on rules and regulations for all European countries:

C1/96: Baltic and Atlantic Coast of Europe
C2/97: Mediterranean and Black Sea

- publishes safety advice:
  C8: Cruising Yacht Safety Sail and Power
- publishes *Cruising News* to update and supplement this information.

The RYA Sail Cruising Committee is made up of elected representatives from all areas of the UK and Northern Ireland. It meets regularly to consider current problems and direct RYA policy in discussions of cruising interest with various authorities in the UK and abroad. The RYA acts as Secretariat to the European Boating Association (EBA). The Association is also responsible for the administration of a superb syllabus of shore-based and sea-based training courses for the cruising yachtsman and yachtswoman.

S

Safe Speed

although there are no speed restrictions on the high seas, the ColRegs (US: NavRules) do define a "safe speed" and this definition has force of law in case of any collision of accident. International Rule 6 reads as follows:

> **Safe Speed**
>
> Every vessel shall at all times proceed at a safe speed so that she can take proper and effective action to avoid collision and be stopped within a distance appropriate to the prevailing circumstances and conditions.
>
> In determining a safe speed the following factors shall be among those taken into account:
>
> a) By all vessels:
> (i) the state of visibility;
> (ii) the traffic density including concentrations of fishing vessels or any other vessels;
> (iii) the manoeuvrability of the vessel with special reference to stopping distance and turning ability in the prevailing conditions;
> (iv) at night the presence of background light such as from shore lights or from back scatter of her own lights;
> (v) the state of wind, sea and current, and the proximity of navigational hazards;

(vi) the draft in relation to the available depth of water.

b) Additionally, by vessels with operational radar:

(i) the characteristics, efficiency and limitations of the radar equipment;

(ii) any constraints imposed by the radar range scale in use;

(iii) the effect on radar detection of the sea state, weather and other sources of interference;

(iv) the possibility that small vessels, ice and other floating objects may not be detected by radar at an adequate range;

(v) the number, location and movement of vessels detected by radar;

(vi) the more exact assessment of the visibility that may be possible when radar is used to determine the range of vessels or other objects in the vicinity.

The only time I am going too fast in a sailing boat seems to be when I'm approaching a berth.

safety harness

a harness used to prevent crew being swept overboard in bad weather conditions; may be improvised from stout cordage but purpose-made webbing harnesses are to be preferred. Worn over the shoulders, around the chest and (advisably) under the crotch and fitted with a ring to which a LANYARD may be connected. The other end of the lanyard is then clipped to a strong point on the boat or to a JACKSTAY (US: jackline). Harnesses can be permanently fitted inside a heavy-weather jacket but the job of getting them in there can be a frustrating problem in three-dimensional geometry. Getting the harness into my Musto Ocean jacket took half a day, even with the help of the instruction book. I also have a safety harness that is integrated with my life jacket—the best arrangement when not wearing oilies/foulies.

safe water marks

a system of BUOYAGE used to indicate an area of safe passage and that may be passed to PORT or STARBOARD: employs a single red ball as a TOPMARK with a red and white color scheme for the main body of the BUOY; lights are OCCULTING white (see IALA). A very jolly buoy that is always a pleasure to see after a long or difficult passage.

sag

to "sag" is to drift off course. Boats hog and sag when the keel is deformed.

sail

1. any arrangement of cloth (traditionally canvas, more usually synthetic materials) arranged in a boat in such a way that it will catch the wind and convert the power of that wind to a motive force; see FORESAIL, GENOA, JIB, LUGSAIL, MIZZEN, MAINSAIL, SPINNAKER, STAYSAIL, TOPSAIL.
2. to "set sail," vernacular expression; to leave, to depart. Originally: to hoist and free the sails to enable them to start pulling.

sail boarding

see WIND SURFING.

sail by

when you "sail by" someone you ignore or snub them, "give them the cold shoulder."

Sailing Alone Around the World

Captain Joshua SLOCUM's classic book of his pioneering 1890s circumnavigation. When this was republished in 1996 (in the UK) I eagerly rushed down to the local book mega-store for my copy. Rather than spend hours trying to find the right section I went to the enquiries desk.

> "Good morning, sir," the assistant smiled.
>
> "Good morning. Where can I find a copy of *Sailing Alone Around the World* by Joshua Slocum?"
>
> "What was the title, again, sir?"
>
> "*Sailing Alone Around the World.*"
>
> "And what's it about, sir?"

See also STOWAWAYS and SLOCUM.

sailing barge

see BARGE.

sailing canoe

a canoe that has wind as a means of propulsion, as well as paddles; usually fitted with OUTRIGGERS. This type of ocean-going canoe is most commonly found in the Pacific.

> The Canoes which were by no means a Master piece of workman Ship, were fitted with outriggers. The Sail was triangular, extended between two sticks one of which was the Mast and the other the Yard or boom, at least they appeared to us who only saw them under Sail at some distance off.
>
> *The Journals of Captain Cook*, Saturday, 27 August 1774

"sailing close to the wind"

BEATING to windward, as near as possible toward the direction of the wind; taking risks.

sailing directions

also known as "pilots," these are, essentially, guidebooks to coastal waters and the safe means of entering ports and restricted waters. Pilots are produced by the HYDROGRAPHER TO THE NAVY (in the UK), private publishers such as Imray Laurie, and sometimes by local sailing clubs. Which to choose? The Admiralty books are comprehensive and cover the whole world. Imray Laurie are equally authoritative, much better illustrated (chartlets and color photographs) but do not attempt to achieve global coverage. For specific, reliable *local* knowledge, sailing club directions are often the answer. They may not be flashily produced, but they are based on the work by the most experienced sailors in that cruising ground. Do not go to sea without at least one type of pilot that covers the coast and ports you are planning to visit. (Pilots do not go far beyond the beach or the pier-head. I also take guidebooks, preferring the *Lonely Planet* or *Rough Guide* series to anything like Baedeker.)

sail into

to attack someone physically or verbally; now a little out of date.

sail plan

the arrangement of sails on a boat; may be a naval architect's drawing.

salon (US)

see SALOON.

saloon

the main CABIN of a boat; on smaller yachts this is also the galley, dining room, library, and part-time bedroom.

salvage

the process and results of recovering a vessel and/or its cargo. Salvage means "to save" and comes from the Latin *salvare*.

sampan

a flat-bottomed boat used on the rivers and in the harbors of China and some other countries in the Far East; traditionally sculled (invariably by a woman as old as Confucius's mother, a cigarette stuck to her lower lip and working a YULOH—SCULLING OAR—over the STERN). I have seen the boats fitted with small inboard engines—the TRANSOM seems too high for an outboard. The name is a distortion of the Chinese *san pan* or "three planks," which implies that the construction was based on a CARVEL-BUILT hull of three planks curving up toward a high BOW. Occasionally fitted with a single JUNK sail.

samson post (or sampson post)

a strong wooden or metal post in the FOREDECK to which the anchor cable (RODE) is attached. May also be found at the STERN for towing. Also cargo-handling DERRICKS used on traditional general-cargo freighters.

schooner

a FORE-AND-AFT rigged sailing boat with two or more masts, the masts being of equal height or the forward mast being shorter than the mainmast or

masts to the STERN. A "TOPSAIL schooner" has one or two SQUARE topsails on the mainmast.

Schooners are as American as apple pie. The first was built in 1713 in the Cape Ann, Massachusetts, fishing port of Gloucester. Perhaps the most famous schooner was the *Hannah*, built at Glover's Wharf not far away at Beverly in 1775 under the patronage of General George Washington. The *Hannah* became the first ship of the US Navy. Another, more infamous schooner was the *America*. In 1851 a team from the New York Yacht Club beat the British team in a race around the Isle of Wight and had off with the trophy. The silverware became known as the "America's Cup" and that is where it stayed for 132 years before Australia and New Zealand began to win it.

There is a lot of doubt about the origin of the word, but I have a theory. It would not be surprising if some skilled Dutch shipwrights worked on the construction of the original. The Dutch word *schoon* means "clean," "beautiful," or "fine." Imagine one of these guys standing back, wiping the sweat from his brow, looking at the near-completed vessel and saying, "Schoon!" From then on they were known as "schooners." If this is not true, it should be.

The schooner was boat of choice for North Atlantic fishing and inshore cargo carrying until the First World War. An American friend once told me that "the schooner is the best thing to come out of America since the British."

Here is Robert Louis Stevenson on schooners, the South Seas, and beer, a perfect combination.

It was about three o'clock of a winter's afternoon in Tai-o-hae, the French capital and port of entry of the Marquesas Islands. The trades blew strong and squally; the surf roared loud on the shingle beach; and the fifty-ton schooner of war, that carries the flag and influence of France about the islands of the cannibal group, rolled at her moorings under Prison Hill. The clouds hung low and black on the surrounding amphitheatre of mountains; rain had fallen earlier in the day, real tropic rain, a waterspout for violence; and the green and gloomy brow of the mountain was still seamed with many silver threads of torrent.

In these hot and healthy islands winter is but a name. The rain had not refreshed, nor could the wind invigorate, the dwellers of Tai-o-hae: away at one end, indeed, the commandant was directing some changes in the residency garden beyond Prison Hill; and the gardeners, being all convicts, had no choice but to continue to obey. All other folks slumbered and took their rest: Vaekehu, the native Queen, in her trim house under the rustling palms; the Tahitian commissary, in his be-flagged official residence; the merchants, in their deserted stores; and even the club-servant in the club, his head fallen forward on the bottle-counter, under the map of the world and the cards of navy officers. In the whole length of the single shoreside street, with its scattered board houses looking to the sea, its grateful shade of palms and green jungle of puraos, no moving figure could be seen. Only, at the end of the rickety pier, that once (in the prosperous days of the American rebellion) was used to groan under the cotton of John Hart, there might have been spied upon a pile of lumber the famous tattooed white man, the living curiosity of Tai-o-hae.

His eyes were open, staring down the bay. He saw the mountains droop, as they approached the entrance, and break down in cliffs: the surf boil white round the two sentinel islets; and between, on the narrow bight of blue horizon, Ua-pu upraise the ghost of her pinnacled mountain-tops. But his mind would take no account of these familiar features; as he dodged in and out along the frontier line of sleep and waking, memory would serve him with broken fragments of the past: brown faces and white, of skipper and shipmate, king and chief, would arise before his mind and vanish; he would recall old voyages, old landfalls in the hour of dawn; he would hear again the drums beat for a man-eating festival; perhaps he would summon up the form of that island princess for the love of whom he had submitted his body to the cruel hands of the tattooer, and now sat on the lumber, at the pier-end of Tai-o-hae, so strange a figure of a European. Or perhaps, from yet further back, sounds and scents of England and his childhood might assail him: the merry clamour of cathedral bells, the broom upon the foreland, the song of the river on the weir.

It is bold water at the mouth of the bay; you can steer a ship about either sentinel, close enough to toss a biscuit on the rocks. Thus it chanced that, as the tattooed man sat dozing and dreaming, he was startled into wakefulness and animation by the appearance of a flying jib beyond the western islet. Two more headsails followed; and before the tattooed man had scrambled to his feet, a topsail schooner, of some hundred tons, had luffed about the sentinel, and was standing up the bay, close-hauled.

The sleeping city awakened by enchantment. Natives appeared upon all sides, hailing each other with the magic cry "Ehippy"—ship; the Queen stepped forth on her verandah, shading her eyes under a hand that was a miracle of the fine art of tattooing; the commandant broke from his domestic convicts and ran into the residency for his glass; the harbour-master, who was also the gaoler, came speeding down the Prison Hill; the seventeen brown Kanakas and the French boatswain's mate, that make up the complement of the war-schooner, crowded on the forward deck; and the various English, Americans, Germans, Poles, Corsicans, and Scots—the merchants and the clerks of Tai-o-hae—deserted their places of business, and gathered, according to invariable custom, on the road before the club.

So quickly did these dozen whites collect, so short are the distances in Tai-o-hae, that they were already exchanging guesses as to the nationality and business of the strange vessel, before she had gone about upon her second board towards the anchorage. A moment after, English colours were broken out at the main truck.

"I told you she was a Johnny Bull—knew it by her headsails," said an evergreen old salt, still qualified (if he could anywhere have found an owner unacquainted with his story) to adorn another quarter-deck and lose another ship.

"She has American lines, anyway," said the astute Scots engineer of the gin-mill; "it's my belief she's a yacht."

"That's it," said the old salt, "a yacht! look at her davits, and the boat over the stern."

"A yacht in your eye!" said a Glasgow voice. "Look at her red ensign! A yacht! not much she isn't!"

"You can close the store, anyway, Tom," observed a gentlemanly German. "*Bon jour, mon prince!*" he added, as a dark, intelligent native cantered by on a neat chestnut. "*Vous allez boire un verre de biere?*"

But Prince Stanila Moanatini, the only reasonably busy human creature on the island, was riding hot-spur to view this morning's landslip on the mountain road; the sun already visibly declined; night was imminent; and if he would avoid the perils of darkness and precipice, and the fear of the dead, the haunters of the jungle, he must for once decline a hospitable invitation. Even had he been minded to alight, it presently appeared there would be difficulty as to the refreshment offered.

"Beer!" cried the Glasgow voice. "No such a thing; I tell you there's only eight bottles in the club! Here's the first time I've seen British colours in this port! and the man that sails under them has got to drink that beer."

The proposal struck the public mind as fair, though far from cheering; for some time back, indeed, the very name of beer had been a sound of sorrow in the club, and the evenings had passed in dolorous computation.

"Here is Havens," said one, as if welcoming a fresh topic.—"What do you think of her, Havens?"

"I don't think," replied Havens, a tall, bland, cool-looking, leisurely Englishman, attired in spotless duck, and deliberately dealing with a cigarette. "I may say I know. She's consigned to me from Auckland by Donald and Edenborough. I am on my way aboard."

"What ship is she?" asked the ancient mariner.

"Haven't an idea," returned Havens. "Some tramp they have chartered."

The Wrecker by Robert Louis Stevenson, 1896

You have to admire this man Havens. He charters a 100-ton topsail schooner for his grand departure from the Marquesas and then describes it as "some tramp."

scope

the length of ANCHOR cable (RODE) deployed; three times the depth of water is the rule of thumb but this can vary according to the amount of chain used (as against rope) the condition of wind and sea; see ANCHORING.

scopolamine

see SEASICKNESS.

screw

see PROPELLER.

scrimshander (or skrimshander)

someone who carves SCRIMSHAW.

scrimshaw (or skrimshaw)

an ornament carved out of whale bone or whale teeth; originally a pastime among American whalers; the origins of the word are obscure and the best guess is that it stems from the name of an early king of the craft:

> "What's this piece of carved whale ivory?"
>
> "That's a scrimshaw. Mr. Scrimshaw makes 'em!"

scrub around, to

from scrubbing wooden decks; to miss something out.

scrubbing decks

Just because you are away at sea does not mean there are no chores to perform—as Richard Henry Dana discovered.

> Tuesday, Sept. 8th [1840]. This was my first day's duty on board the ship; and though a sailor's life is a sailor's life wherever it may be, yet I found everything very different here from the customs of the brig Pilgrim. After all hands were called, at daybreak, three minutes and a half were allowed for every man to dress and come on deck, and if any were longer than that, they were sure to be overhauled by the mate, who was always on deck, and making himself heard all over the ship. The head-pump was then rigged, and the decks washed down by the second and third mates; the chief mate walking the quarter-deck and keeping a general supervision, but not deigning to touch a bucket or a brush. Inside and out, fore and aft, upper deck and between decks, steerage and forecastle, rail, bulwarks, and water-ways, were washed, scrubbed and scraped with brooms and canvas, and the decks were wet and sanded all over, and then holystoned. The holystone is a large, soft stone, smooth on the bottom, with long ropes attached to each end, by which the crew keep it sliding fore and aft, over the wet, sanded decks. Smaller hand-stones, which the sailors call "prayer-books," are used to scrub in among the crevices and narrow places, where the large holystone will not go. An hour or two, we were kept at this work, when the head-pump was manned, and all the sand washed off the decks and sides. Then came swabs and squilgees; and after the decks were dry, each one went to his particular morning job. There were five boats belonging to the ship,—launch, pinnace, jolly-boat, larboard quarter-boat, and gig—each of which had a coxswain, who had charge of it, and was answerable for the order and cleanness of it. The rest of the cleaning was divided among the crew; one having the brass and composition work about the capstan; another the bell, which was of brass, and

> kept as bright as a gilt button; a third, the harness-cask; another, the man-rope stanchions; others, the steps of the forecastle and hatchways, which were hauled up and holystoned. Each of these jobs must be finished before breakfast; and, in the meantime, the rest of the crew filled the scuttle-butt, and the cook scraped his kids (wooden tubs out of which the sailors eat) and polished the hoops, and placed them before the galley, to await inspection. When the decks were dry, the lord paramount made his appearance on the quarter-deck, and took a few turns, when eight bells were struck, and all hands went to breakfast.
>
> *Two Years Before the Mast* by Richard Henry Dana, 1840

It is a lot easier with GRP.

scud, to

to run before the wind in gale or storm conditions under reduced sail or bare poles; no longer recommended best practice (if it ever was)—see STORM TACTICS.

sculling oar (scull, to)

usually refers to a long oar used over the **STERN** and moved from side-to-side to propel a boat through the water; can also be used as a **JURY-RIGGED** rudder. The motion of the sculling oar tends to have an opposite effect on the hull of the boat being sculled, but as the movement is angled, the hull goes forward as well as sideways. See **YULOH**.

scuppers

drain holes in the **BULWARKS** that allow seas coming over the side to wash away from the **DECK**; if your boat is sliding down the face of a wave they can actually let large gushes of water *in*.

scurvy

a disease caused by a deficiency of vitamin C.

The problem with scurvy is well known but at the time of George Anson's 1740 expeditionary voyage, its origins were still a mystery. During the 18th century, scurvy killed more British sailors than enemy action. It was to this disease that Anson lost two-thirds of his crew of 1900 in the first ten months of the voyage. During the Seven Years' War (1756–1763), the Royal Navy conscripted 184,899 sailors, of whom 133,708 were lost, mostly to scurvy; an appalling rate of attrition.

Initial symptoms include lethargy, rashes on the legs, spongy gums, and bleeding from the mouth. If untreated it can and will lead to fever, jaundice, festering wounds, neuropathy (damage to nerve endings), and death.

An effective treatment, the consumption of citrus fruit or fresh vegetables, was formally documented by Royal Navy surgeon James Lind (James Lind, *A Treatise on the Scurvy* [London: A Millar, 1753]) but known to observant seafarers before that. And to Mrs. Ebot Mitchell of Gloucestershire, England. As long ago as 1707 Mrs. Mitchell wrote in her notebook a "Recp.t for the Scurvy." This consisted of extracts from fresh vegetables mixed with a "plentiful supply of orange juice, white wine or beer." However, many naval officers believed the cause was idleness (probably because lethargy is a symptom) and the notoriously conservative Admiralty took no action until later that century.

The link to vitamin C wasn't identified until 1932. But in the meantime adding lime juice to the ship's ration of GROG kept the scurvy problem at bay and invented the cocktail.

scuttle, to (to scupper)

to scuttle a ship, yacht, or whatever, is to deliberately make it sink; usually done by opening the SEACOCKS but on larger vessels an explosive charge is likely to be more effective. Take your revenge by planting the Semtex behind the HEADS—but get well clear—especially if you have a holding tank. The whole process is very sad, but its purpose is usually to remove a hazard from shipping lanes. (To scupper means the same thing.)

> I entertained on a cruising trip that was so much fun that I had to sink my yacht to make my guests go home.
>
> F. Scott Fitzgerald

HMS *Britannia* was scuttled after the death of her owner, King George V.

scuttlebutt (US)

the gossip or a gossip; named after a water cask (usually at the bottom of the main COMPANIONWAY) around which sailors used to gather and drink; the nautical equivalent of the office coffee machine.

scuttles

the original but now-obsolete name for what everyone calls a porthole or PORT LIGHT.

sea anchor

just about anything fixed to a stout line, lashed to a strong point and thrown over the BOW in order to slow the vessel down and keep the bow heading toward the wind in severe weather conditions. Can be improvised from anything that will cause drag (e.g., a truck tire), or purpose-built such as a DROGUE. The latest solution is to use a military specification parachute. See STORM TACTICS.

seabed (or sea bed)

the bottom of the sea; the place you want your KEEL to avoid and your ANCHOR to get married to.

sea breeze

see BREEZE, LAND AND SEA.

sea cock

a valve in any pipe passing through the hull of a vessel that prevents the ingress of sea water.

sea-kindly

see SEAWORTHY.

sea legs

getting used to the movement of a boat or ship when moving about the vessel. This is what the redoubtable but miserable Miss Emily Brittle wrote to her parents after they sent her on "a trip to the Indies . . . in pursuit of a swain" (boyfriend).

> With tossing and tumbling my bones were so sore,
> Such an up and down motion I ne'er felt before;
> Many days had elapsed e'er I first got the notion
> That to keep on my legs I must humour the motion.

from *The India Guide; or, Journal of a Voyage to the East Indies in the Year MDCCLXXX*
by Sir George Dallas (1758–1833)

MDCCLXXX is 1780 and another good reason why roman numerals should have died with the Roman Empire. (After writing that remark in the first edition I received a note from a prominent politician saying, "Not on my watch!") For more on Miss Brittle "humouring the motion," see FIDDLE.

sea mark

a navigation mark on the sea (e.g., a BUOY); a navigation mark on the land (or, indeed, a mark on land used for navigation) is called a LAND MARK.

sea room

navigable distance from the shore or any other obstacle.

seasickness (*mal de mer*)

a form of motion sickness usually brought on by a feeling of well-being and euphoria. Reputedly comes in two stages: during the first you fear you might die; during the second, you fear you might not. This is what happened to Robinson Crusoe on his very first day at sea. He left from the Yorkshire port of Hull on the north bank of the River Humber.

> The ship was no sooner gotten out of the Humber, but the wind began to blow, and the waves to rise in a most frightful manner; and as I had never been at sea before, I was most inexpressibly sick in body, and terrified in my mind. I began now seriously to reflect upon what I had done, and how justly I was overtaken by the judgment of Heaven for my wicked leaving my father's house, and abandoning my duty; all the good counsel of my parents, my father's tears and my mother's entreaties, came now fresh into my mind, and my conscience, which was not yet come to the pitch of hardness which it has been since, reproached me with the contempt of advice and the breach of my duty to God and my father.
>
> All this while the storm increased, and the sea, which I had never been upon before, went very high, though nothing like what I have seen many times since; no, nor like what I saw a few days after. But it was enough to affect me then, who was but a young sailor, and had never known anything of the matter. I expected every wave would have swallowed us up, and that every time the ship fell down, as I thought, in the trough or hollow of the sea, we should never rise more; and in this agony of mind I made many vows of resolutions, that if it would please God here to spare my life this one voyage, if ever I got once my foot upon dry land again, I would go directly home to my father, and never set it into a ship again while I lived; that I would take his advice, and never run myself into such miseries as these any more. Now I saw plainly the goodness of his

> observations about the middle station of life, how easy, how comfortably he had lived all his days, and never had been exposed to tempests at sea, or troubles on shore; and I resolved that I would, like a true repenting prodigal, go home to my father.
>
> These wise and sober thoughts continued all the while the storm continued, and indeed some time after; but the next day the wind was abated and the sea calmer, and I began to be a little inured to it. However, I was very grave for all that day, being also a little sea-sick still; but towards night the weather cleared up, the wind was quite over, and a charming fine evening followed; the sun went down perfectly clear, and rose so the next morning; and having little or no wind, and a smooth sea, the sun shining upon it, the sight was, as I thought, the most delightful that ever I saw.
>
> *Robinson Crusoe* by Daniel Defoe, 1719

Many other famous sailors have suffered from seasickness including Admiral Horatio Nelson and pioneering circumnavigator Sir Francis Chichester. Nelson had it so bad it made his arm fall off (this lie is intended to make you feel a little better the next time you are afflicted and your arms are the only parts of you that feel fine).

> CAPTAIN. I'm never, never sick at sea!
>
> ALL. What never?
>
> CAPTAIN. No, never!
>
> ALL. What, never?
>
> CAPTAIN. Hardly ever!
>
> from *HMS Pinafore* by Gilbert & Sullivan

For the best insight try the chapter on seasickness in *Psychology of Sailing* by Prof. Michael Stadler.

seasickness remedies

prevention is better than cure, but once the onset has started, the following may help.

- The ultimate, unfailing remedy is to stand under a tree. Anchor in a quiet bay. Go ashore.
- Lie down and go to sleep.
- Ginger, crystallized ginger, ginger cookies. Recommended, but I have never tried it myself. All my ginger goes in the curries. In the UK, by the way, one

company sells ginger biscuits under the brand name "Ginger Nuts." I doubt this is true of any American manufacturers.
- I know that melon works because I have tried it on others. Not only is it easy to consume, it has a non-intrusive flavor. It is also 90 percent water so it deals with the dehydration problem at the same time.
- A hot water bottle on the tummy under the foul weather gear. If no hot water bottle is to hand, try the dog, should you have one onboard.
- If it is the fumes from the engine that are getting to you, try a little Tiger Balm around the nostrils. If the Tiger Balm makes you feel sick, I am out of ideas.
- Wristbands—worn on both arms—that press a bead on to an acupressure point between the two veins in your inner wrist. I suspect these work if you really believe they will work.
- Patches that you stick on your upper arm.
- Scotch tape (UK: Sellotape) that you stick across your navel.
- Spectacles that incorporate an artificial horizon. (Honestly! They are made by a company in Belgium and tested by the French Navy.)
- Cinnarizine. An antihistamine anti-emetic sold under the brand names Stugeron, Stugeron Forte, Cinaziere, Marzine-RF. The main side effects are drowsiness and lethargy (but never as bad as that brought on by the dreaded *mal* itself). Does not combine well with alcohol. Expensive, but widely available and effective for most people. Not really a remedy, more a preventative. Take these about an hour before setting sail—once the first "pizza" is on the deck, it is too late. I have seen it suggested that you should take a couple of the tablets before going to bed, then two more over breakfast for maximum effectiveness. A side benefit of this particular idea is that you will be able to walk on the water to get to the mooring; I don't try climbing the rigging though.
- Meclozine. Another remedy sold under the brand name of Sea Legs.

Finally, it does go away. It might take some days, but you are going to be fine. See DEHYDRATION.

seasickness; skipper's guidelines

Sickness ruins everyone's seagoing experience. Therefore it is in everyone's interest for the skipper to exercise his or her leadership and management skills to avoid it. Here is a six-point plan of action.

1. **Reduce anxiety by briefing beginners on what to expect.** Explain that a sailing boat is supposed to lean over and just because some water might come over the side does not mean that you are about to sink. (This part

tends to be countered by the safety briefing that you are going to give next. The very presence of life jackets, a liferaft, flares, and an EPIRB tends to counter all that "you'll be safe with me" stuff.)

2. **Get everyone to take pills.** To encourage the macho types who don't know what they're in for, take them yourself—even if you don't get sick. A little prestidigitation may make it unnecessary to swallow the things for real; some types make you drowsy.
3. **Keep everyone busy.** Make sure that everyone has something to do on deck, in the fresh air. Put the first person who goes suspiciously quiet on the helm. If you have a second victim, put them on "look-out" duty (even if you are mid-Atlantic). Threaten to throw them overboard if they do not do a double 360° scan of the horizon every few minutes.
4. **Try to make it a positive experience.** The best possible weather in the most exotic location helps considerably. Keep the first leg as short as possible and do not cook. Prepare sandwiches and salads and be ready to make instant soups, noodles, and so forth for anyone who gets hungry. (This kind of planning is not easy. You may decide to sit out a blow in the marina and entertain your guests in the pub but, although the wind may have dropped the following day, you may still encounter the aftermath in the form of an uncomfortable swell.)
5. **Do avoid starting the engine until essential.** Diesel fumes are a killer for some people.
6. **Be prepared to stop.** Once the first victim has been reacquainted with their breakfast, others will go down like ninepins. Head in to the nearest sheltered bay or haven. Going ashore is the fastest cure I know.

Also, skippers should carefully monitor the condition of victims. If the sickness persists over a long period and the patient is unable to keep down any fluids then DEHYDRATION will set in.

seaworthy

when a boat is able to handle rough weather she is said to be "seaworthy" or "sea-kindly."

secondary port (UK)

equivalent to a SUBORDINATE STATION in US tide tables; an intermediate point between STANDARD PORTS for which an additional calculation has to be made.

sedan cruiser (US)

a motor cruiser with a SALON and a raised bridge.

seized

bound up, unable to come apart with the use of exceptional force (e.g., the use of a good hammer and a very bad temper on the end of the spanner (wrench).

seizing, a

a binding together of the two ends of a LINE.

selective availability

see GLOBAL POSITIONING SYSTEM.

self-steering gear

see VANE SELF-STEERING GEAR, HYDRAULIC SELF-STEERING GEAR.

self-tending (self-tacking)

descriptive of a sailing boat that automatically re-SHEETS the sails for the opposite TACK as the vessel goes about.

semi-displacement hull

a hull with soft CHINES or a gradually rounded bottom that just about allows a boat to PLANE.

serve, to

the process of binding a rope with twine or adhesive tape.

set

1. the direction of a tidal current; unlike the direction of the wind, this is always expressed in terms of the direction in which the current is flowing. A westerly wind, therefore, is one that comes *from* 270°. However, a current coming from the same direction is expressed as having a set of 90°. I have no rational explanation for this, but it is crucial that the navigator gets it right.
2. to “set” a sail is to hoist it into its working position.

seventh wave

the seventh in a series of WAVES is said to the be worst, always the "killer" in a storm. This is no more than superstition, a nautical equivalent of an urban myth.

sewn up

it is "all sewn up" for you when they stitch up your HAMMOCK (with you and a spare cannonball in it) before slipping you overboard and into the BRINEY. Note that international regulations now require commercial yachts to carry BODY BAGS—which must be brought to port full if used. Solo sailors have a problem here. Will they have committed an offense under Maritime Law if they fail to die inside the ship's statutory body bag? You may need to know this—you would not want to die a criminal.

sextant

traditional navigation instrument used for measuring the angle between two objects; used both horizontally and vertically in COASTAL NAVIGATION, vertically in OCEAN NAVIGATION to measure the angle above the horizon of CELESTIAL OBJECTS. So-called because it subtends one-sixth of a circle (60°). A development of the quadrant and the ASTROLABE.

shackle, snap shackle

a U-shaped metal link that can be opened and closed for connecting line to secure rings, chain to ANCHOR, HALYARDS to the heads of sails, and so forth. The basic form of shackle has a pin that reaches through one hole at the end of an arm and screws into a threaded hole at the end of the opposite arm. Another arrangement has the pin hinged on one arm so that it is folded down to engage the other; this is known as a snap shackle. Shackles are used extensively on sailing boats.

Shadwell Prize

For many centuries the British Hydrographic Office relied upon Royal Navy officers to survey distant ports, anchorages, and coastlines for inclusion in the charts and pilots they published. If you look at the bottom right corner of British charts and the sketches in pilots (sailing directions) you will often see a credit to the work of some officer in the 18th or 19th century.

This tradition is probably less important in the era of satellite imagery, but it seems that it hasn't been forgotten. In December 2002, Commander Tom Tulloch of the Royal Canadian Navy was awarded the Hydrographic Office's Shadwell Prize for his work in surveying little-known harbors and potential terrorist bolt-holes in the eastern Mediterranean, the Red Sea, the Arabian Sea, and the Iranian (or Persian) Gulf. During two tours between January 2001 and March 2002 he systematically surveyed every port entered by his ship—and had the ship's photographer take photographs. Cdr. Tulloch then filed his reports with the British Hydrographic Office (the Canadian Hydrographic Office only covers Canadian waters). In return, he was awarded the "Shadwell Testimonial Prize," which was established in 1888. Remarkably, this is the second time he has won the prize—a unique achievement.

See HYDROGRAPHER OF THE NAVY.

shake out, to

to let out a REEF.

shallow, shallows

see SHOAL.

Shanghai

to forcibly recruit someone to the marine (usually through a combination of drink, drugs, and the odd blow over the head with a BELAYING PIN); a great old Royal Navy tradition taken up by many other nations. Shanghai itself is a fascinating city. In Chinese *shang* is used to signify the start of something and *hai* means "sea"; so, Shanghai is on the Huangpu River just south of where it joins the estuary of the magnificent Yangzi.

shank

the main shaft of any ANCHOR with the connecting ring at one end and the FLUKES at the other.

sheath

1. the outer part of a modern rope comprising a CORE and a COVER of synthetic material.
2. a convenient means of ensuring that small electrical and electronic components are kept dry and free of saltwater contamination; thought to have various other applications; available at local pharmacies and the men's room in pubs and bars.

sheave

the revolving wheel(s) in a BLOCK; also called PULLEYS.

sheer

1. to sheer: a vessel at ANCHOR sheers when she swings from side to side from the effect of wind and tide.
2. the sheer of a boat is the exterior line of the deck from BOW to STERN.

sheet

the LINES connected to the CLEW of a sail that enable its angle to the FORE-AND-AFT line of the vessel (and, therefore to the wind) to be adjusted for optimum performance. "Three sheets to the wind" originally meant that a vessel was over-canvased and had gone out of control. Now descriptive of someone who spent too long in a dockside bar, has lost the flashlight, and is rowing the inflatable around the moorings trying to find his yacht.

ship fever

epidemic typhus (not to be confused with typhoid fever, something completely different).

Ill-health was the major cause of attrition on ships in the age of sail. Of the 1,900 men who boarded the ships of Commander Anson's SQUADRON in 1740, only 188 made it home. The majority of those lost had died of malnutrition or disease—especially SCURVY.

The consequences of scurvy were well known but at the time of Anson's voyage its causes were still a mystery. And scurvy wasn't the beginning and end of the story. Bad food leads to rampant dysentery and some of the ships may have been forced to rig extra THUNDER BOXES. The real killer though was typhus, a disease carried by body lice that flourished in the humid and insanitary conditions on Anson's vessels. The contemporaneous name for typhus was "ship fever."

When the squadron anchored off the Brazilian island of Santa Catarina on 21 December 1740 the Commodore's first action was to order all the sick to be sent ashore. The locals must have been less than delighted. The ships were then scrubbed clean by the remaining crew who then lit fires below deck and sealed off the hatches and gun ports in an attempt to fumigate the rats, lice, and fleas. A final wipe-down with vinegar-soaked rags was supposed to be the *coup de grace*; but I have my doubts.

Respite on "the hard" was probably too late for the most ill. Of 80 men who went ashore from HMS *Centurion*, 26 died. But all crew members were living in makeshift tents pitched on the mosquito-infested shore and many of them caught malaria. When the *Centurion* hauled anchor on 18 January 1741 she still had 96 sick crewmembers onboard.

shipping lanes

traditionally used to refer to any stretch of water commonly used by passenger and cargo ships in passage between two ports. In coastal waters, such "lanes" are often designated on charts and may be part of a TRAFFIC SEPARATION SCHEME.

ship in seas, to

to take in seas—to be flooded by a wave.

ship's boy

See MIDSHIPMAN.

ship versus boat

When is a ship not a ship? And when is it a boat? "A ship is bigger than a boat" is all that most seafarers will agree on, but many other definitions have been postulated.

- A ship does not move when you climb onboard (a boat does—unless it is securely chocked on the hard).
- A ship is a vessel that carries smaller craft (boats) in DAVITS on a boat deck. That fits well with the US Navy's definition of a "boat" being a vessel small enough to be hoisted onto a bigger vessel—known as a "ship." Therefore, a

submarine cannot be a ship because it does not carry boats in davits on its topsides and, sure enough, the US Navy insists that all submarines are boats. So, a submarine must be a boat, even though a big one does not move when you step aboard.
- A destroyer will carry boats in davits, thus making it a ship. But, if you put the destroyer on the flight-deck of an aircraft carrier, would that make it a boat? Probably not.
- A boat is traditionally defined as a craft used to transfer officers and crew between ship and shore. Not something a destroyer is used for. Nor a submarine.

Is that clear? All right, it might be difficult to define a ship or a boat, but most sailors certainly know a ship when they see one bearing down on them.

shipwreck

the loss of a ship at sea or, more usually, on a hostile coastline or REEF.

"Shipwreck" is a headline word that conveyed much more horror in the 19th century when travelers commonly had to face the perils of a voyage on the high seas. In those days, bad weather, lee shores, and unsound ships conspired to present a much greater threat to life and limb than the likelihood of a madman with a mission spoiling the relative comfort and safety of a few hours on a modern jet liner. Also lost was a style of *reportage* used to describe the terror of a shipload of men, women, and children fetching up on the rocks. Here is a report on the loss of the *New Era* on Deal Beach, north of the port of Dover in the English Channel in 1855.

APPALLING SHIPWRECK OF THE *NEW ERA*

We are pained to record one of the most fearful disasters that has ever taken place on our coast, in the wreck of the new ship *New Era*, freighted with 427 passengers from Bremen to New York. On the morning of November 13th, after a passage of 46 days, the *New Era* neared our coast, where, in thick weather, the soundings alone furnish the only tangible evidence of a vessel's position. At the call of the morning watch, the Captain, after attending to the cast of the lead, retired to his cabin, leaving the second mate in charge of the deck. Added to the discomforts of a lengthy passage, the ship proved leaky, and the working of the pumps had been apportioned as a part of the duties of both passengers and crew. On the last night of the fatal voyage the wind blew a stiff breeze from S.E., which caused considerable sea, and the ship being enveloped in fog, which was scarcely illumined by the dawn of day, when the fearful echo of *breakers* ran through the crowded decks of the doomed vessel—and before six o'clock the ship struck on Deal Beach, swung broad-side to, and as she settled in the sand the sea made a clear breach over her; a few feeble and abortive efforts were made to get a line to the shore, and failing in this, by means of the boats, the Captain, officers and most of the crew escaped to the shore, and six hours after stranding the deserted ship had no commander, or a single man on board who understood what was being done on shore for the assistance of the unfortunate passengers, none of whom could speak the English language.

On the following morning after every living person had been rescued from the ship, only 143 (including the crew) of the 427 embarked at Bremen, were found to have escaped; making a loss of 284 lives.

A more frightful loss of life on ship-board has scarcely ever been recoded in the annals of emigrant voyages, reckless as these are sometimes made. Such a shameful neglect of the commonest precautions on approaching the coast, and the subsequent desertion of the helpless passengers, calls for a searching inquiry into the loose conduct and inhumanity of those in charge of the fated ship. It is a grave question of most significant import, whether there shall, or shall not be (as at present), a remedy for such culpable recklessness as this that consigns the thrusting passenger to the tender mercies of *fate*, whenever he sets foot on ship-board. It is notorious that emigrants are landed here safer from their own vessels than from ours, in many cases. We do not hesitate to say, that some one should be made responsible for the safe termination of a voyage by *sea*, in the same manner that land conveyancers are held for the safety of life and limb. Too many abuses are arising from the common loose construction of that stereotyped proviso found in all bills of lading, viz., "*the dangers of the seas excepted,*" to permit this time-worn maxim in marine risks to pass unchallenged by scrutinizing investigation. The public mind is fast becoming awakened, and will not fail to adopt some measure for securing the better protection of human life, when it is intrusted to the skill of the nautical mechanic or the fidelity of the mariner. A far greater degree of responsibility must be imposed somewhere.

The Monthly Nautical Magazine, and Quarterly Review, April–September 1855

The shipwrecks that make today's television news headlines seem to fall into two categories: ferries that are badly designed (Europe) or overloaded (Asia and Africa); and bulk carriers that are too big in relation to their structural integrity. In the latter case the consequent threat to the marine environment is compounded by the vessels having single hulls, single engines, and too many venally incompetent skippers. Writing in 2013, I can probably add cruise ships to this list.

shoal

a shallow area of water. The words "shoal" and "shallow" both originate from an Old English word *sceald*, which, unsurprisingly, means shallow.

shoot

1. to make a little distance to windward in a yacht by suddenly LUFFING up when at speed; very useful skill when RACING.
2. to measure the elevation of a CELESTIAL OBJECT using a SEXTANT.

shoreline

where the sea meets the land.

shroud

a wire running from the DECK edge or gunwhale up to the mast as a means of providing lateral support; most yachts have one shroud each side, some two; see also SPREADERS. On BERMUDAN RIGS the shrouds, the FORESTAY, and the BACKSTAY serve to keep the mainmast absolutely rigid. However, if any of these stays fail, the mast could (and probably will) collapse immediately and with catastrophic results. If you ever wondered why heavy-duty bolt crops are recommended equipment onboard all Bermudan-rigged sailing boats, now you know.

> A mariner sat in the shrouds one night,
>
> The wind was piping free;
>
> Now bright, now dimmed, was the moonlight pale,
>
> And the phospher gleamed in the wake of the whale,
>
> As it floundered in the sea.
>
> Elizabeth Oakes Smith

The mariner was more likely to be sitting in the RATLINES or on a CROSSTREE.

sidelights

red and green navigation lights required by the ColRegs (US: NavRules); may, in practice, be fitted at the beam, at the bow, or at the masthead. The following rules apply.

> Rule 21 (b) "Sidelights" means a green light on the starboard side and a red light on the port side each showing an unbroken light over an arc of the horizon of 112.5 degrees and so fixed as to show the light from right ahead to 22.5 degrees abaft the beam on its respective side. In a vessel of less than 20 meters in length the sidelights may be combined in one lantern carried on the fore and aft centerline of the vessel.
>
> Rule 21 (c) "Sternlight" means a white light placed as nearly as practicable at the stern showing an unbroken light over an arc of the horizon of 135 degrees and so fixed as to show the light 67.5 degrees from right aft on each side of the vessel.
>
> **International Rule 25: Lights and Shapes**
>
> Sailing Vessels Underway and Vessels Under Oars
>
> 1) A sailing vessel underway shall exhibit:
>   (i) sidelights;
>   (ii) a sternlight.
> 2) In a sailing vessel of less than 20 meters in length the lights prescribed in paragraph (a) of this Rule may be combined in one lantern carried at or near the top of the mast where it can best be seen.
> 3) A sailing vessel underway may, in addition to the lights prescribed in paragraph (a) of this Rule, exhibit at or near the top of the mast, where they can best be seen, two all-round lights in a vertical line, the upper being red and the lower green, but these lights shall not be exhibited in conjunction with the combined lantern permitted by paragraph (b) of this Rule.
> 4)
>   (i) A sailing vessel of less than 7 meters in length shall, if practicable, exhibit the lights prescribed in paragraph (a) or (b) of this Rule, but if she does not, she shall have ready at hand an electric torch or lighted lantern showing a white light which shall be exhibited in sufficient time to prevent collision.
>   (ii) A vessel under oars may exhibit the lights prescribed in this Rule for sailing vessels, but if she does not, she shall have ready at hand an electric torch or lighted lantern showing a white light which shall be exhibited in sufficient time to prevent collision.

5) A vessel proceeding under sail when also being propelled by machinery shall exhibit forward where it can best be seen a conical shape, apex downwards.

See also ANCHOR LIGHT.

sidereal compass

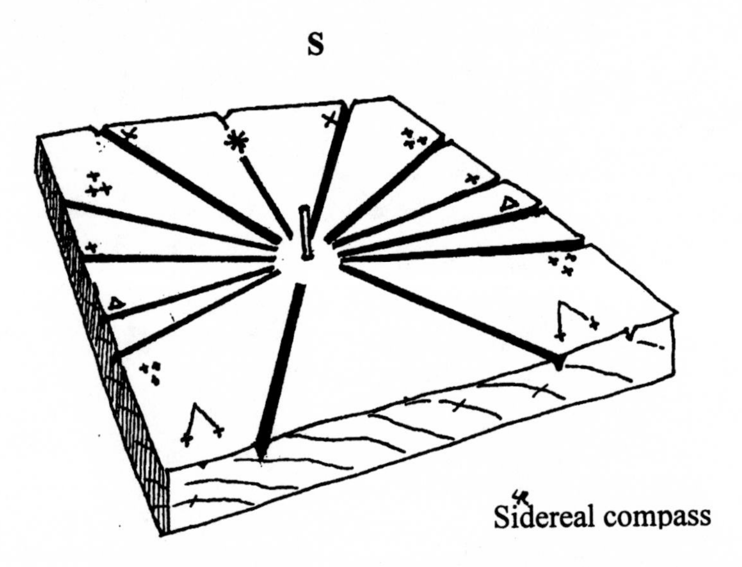

a compass that uses stars to determine orientation; in the Northern Hemisphere, for example, the "top" of the compass can be pointed toward Polaris to provide North and the angle where other stars rise and fall will provide additional "POINTS."

sight

see SHOOT.

signal flags

a universal system of signaling between ships developed long before the advent of VHF; the flags themselves are classic examples of good design combining a limited number of primary colors and simple bold shapes to great effect.

> We arrived at Ilha Grande, our destination, on the 7th day of January, 1887, and came to anchor in nine fathoms of water, at about noon, within musket-range of the guard-ship, and within speaking distance of several vessels riding quarantine, with more or less

> communication going on among them all, through flags. Several ships, chafing under the restraint of quarantine, were "firing signals" at the guard-ship. One Scandinavian, I remember, asked if he might be permitted to communicate by cable with his owners in Christiana. The guard gave him, as the Irishman said, "an evasive answer," so the cablegram, I suppose, laid over. Another wanted police assistance; a third wished to know if he could get fresh provisions—ten milreis' ($5) worth (he was a German)—naming a dozen or more articles that he wished for, "and the balance in onions!" Altogether, the young fellows on the guard-ship were having, one might say, a signal practice.
>
> *Voyage of the "Liberdade"* by Capt. Joshua Slocum, 1894

Today signal flags are mostly used for their secondary meanings; U for "You are standing into danger," A for "Diver down," H for "Pilot onboard" and so on. All this is specified as part of the International Code of Signals.

single sideband (SSB) radio

a shortwave radio used for long distance voice communications and access to weather fax on ocean-going vessels; see also VHF RADIO.

Sirocco (French), Scirocco (Italian)

a southerly wind that blows across the Sahara Desert, then north over the western part of the Mediterranean Sea and into southern continental Europe. Can blow for days and weeks in the summer months. Hot and uncomfortable. The word originates from the Arabic *sharuq*, meaning "east." Blows a lot of dust on DECK and reduces visibility.

skeg, skegg

an extension of the KEEL used to protect the RUDDER.

skiff

a small traditional boat with a flat bottom and, therefore, a very shallow DRAFT; can be rowed, sailed, or have an OUTBOARD ENGINE clipped to the TRANSOM. Young Jack London taught himself to sail on a skiff. But it was with some trepidation that he invited an experienced merchant mariner onboard for a trip across San Francisco Bay.

> He was a runaway English sailor. I was a lad of twelve, with a decked-over, fourteen-foot, centre-board skiff which I had taught

> myself to sail. I sat at his feet as at the feet of a god, while he discoursed of strange lands and peoples, deeds of violence, and hair-raising gales at sea. Then, one day, I took him for a sail. With all the trepidation of the veriest little amateur, I hoisted sail and got under way. Here was a man, looking on critically, I was sure, who knew more in one second about boats and the water than I could ever know. After an interval, in which I exceeded myself, he took the tiller and the sheet. I sat on the little thwart amidships, open-mouthed, prepared to learn what real sailing was. My mouth remained open, for I learned what a real sailor was in a small boat. He couldn't trim the sheet to save himself, he nearly capsized several times in squalls, and, once again, by blunderingly jibing over; he didn't know what a centre-board was for, nor did he know that in running a boat before the wind one must sit in the middle instead of on the side; and finally, when we came back to the wharf, he ran the skiff in full tilt, shattering her nose and carrying away the mast-step. And yet he was a really truly sailor fresh from the vasty deep.
>
> "Small-Boat Sailing" from *The Human Drift*
> by Jack London, 1911

Of course, the "experienced merchant mariner" might have been a cook.

skipjack

1. a type of TUNA; sometimes called bonito.
2. a HARD-CHINE SLOOP-rigged American work boat. Introduced to Chesapeake Bay in 1860, the skipjack incorporated an interesting innovation: its mast raked back so it (and the HALYARD) could be used for cargo handling in port.

skipper

anyone in charge of vessel; from the Dutch *schipper* or "shipper."

slack water

a period of minimum tidal flow as a tide "turns"; lasts about 30 minutes at high and low water.

slant

when a seaman adjusted the HEEL of a ship in order to improve comfort and/or efficiency, he was said to be "putting a new slant on things." (Does this make a "spin doctor" a "slant doctor" perhaps?)

slatting

the (often violent) flogging or flapping of the LEECH of a sail.

slats

see BATTENS.

sleep

It is important to try and get plenty of this when passage making.

> With lack of sleep and too much understanding I grow a little crazy, I think, like all men at sea who live too close to each other and too close thereby to all that is monstrous under the sun and moon.
>
> from *Rites of Passage* by Sir William Golding, 1980

slick

a "slick" is a smooth area of water left by the hull of a vessel being driven BEAM-ON by the wind; also the area to WINDWARD if in very rough weather and you are HOVE TO or running under bare poles or lying a-hull, if you leak oil from a container.

slip

1. a hard, sloping ramp that runs into the water (at least at high tide) thus enabling it to be used for the launching and recovery of boats.
2. a parking space for a boat between two floating piers.
3. the loss of power delivery as a propeller spins in the water; something like a wheel-spin in a car.
4. when an ANCHOR cannot be recovered (or you do not want to recover it), it can be "slipped" letting it all go to the bitter end. Unless you have a diver onboard, it is highly recommended to attach a buoy to the inboard end of the chain (RODE) for ease of recovery.

Slocum, Captain Joshua

"America's best known sailor" was born on 20 February 1844 in Mount Hanly, Nova Scotia—giving rise to claims that he is really "Canada's best known sailor." Actually, in 1844, Nova Scotia was a directly ruled British colony and only in 1867 did it become part of the Canadian Confederation. So perhaps Slocum is "Nova Scotia's best known sailor"?

At the age of sixteen Joshua ran away to sea which, in those days of sail, was certainly a more arduous and risky undertaking than becoming a flight attendant. He learned seamanship and navigation and, by the age of twenty-five, was a merchant marine captain. Eventually he was to own his own **BARQUE**, the *Aquidneck*. Slocum certainly seems to have led an adventurous life but I doubt somehow that it was more or less adventurous than that of any other skipper plying his trade in far-flung corners of the world in the 19th century. What makes it seem exceptional was that Slocum wrote about his experiences; and the impression we have today (I include myself in this) is very much from his books.

Reviews and references to his achievements usually border on the adulatory, but recently I have seen comments that are close to being off-message. These include suggestions that Slocum was a violent man and in one incident killed a mutineer. In 1887 the *Aquidneck* was trading between Rio de Janeiro and Buenos Aires and her skipper was having crewing problems. Against his better judgment Slocum hired four dubious characters in the Brazilian port.

> The ringleader of the gang was a burly scoundrel, whose boast was that he had "licked" both the mate and second mate of the last vessel he had sailed in, and had "busted the captain in the jaw" when they landed in Rio, where the vessel was bound, and where, of course, the captain had discharged him. . . . His chum and bosom friend had come pretty straight from Palermo penitentiary at Buenos Aires when he shipped with me at Rosario.
>
> It was no secret on board the bark that he had served two years for robbing, and cutting a ranchman's throat from ear to ear. These records, which each seemed to glory in, were verified in both cases.
>
> I met the captain afterwards who had been "busted in the jaw"—Captain Roberts, of Baltimore, a quiet gentleman, with no evil in his heart for any one, and a man, like myself, well along in years.
>
> Two of the gang, old Rosario hands, had served for the lesser offence of robbery alone—they brought up in the rear! . . .
>
> The ringleader bully had made unusual efforts to create a row when I came on board early in the evening [of 23 July]; however, as he had evidently been drinking, I passed it off as best I could for the natural consequence of rum, and ordered him forward; instead of doing as he was bid, when I turned to hand my wife to the cabin he followed me threateningly to the break of the poop. What struck me most, however, was the conduct of his chum, who was sober, but in a very unusual, high, gleeful mood. It was knock-off time when I came along to where he was seizing off the mizzen topgallant backstay, the last of the work of refitting the late *pampeiro* damage; and the mate being elsewhere engaged, I gave the usual order to quit

work. "Knock off," I said to the man, "and put away your tools. The bark's rigging looks well," I added, "and if to-morrow turns out fine, all will be finished"; whereupon the fellow laughed impertinently in my face, repeating my words, "All will be finished!" under his breath, adding, "before tomorrow!" This was the first insult offered by the "Bloodthirsty Tommy," who had committed murder only a short time before; but I had been watched by the fellow, with a cat-like eye at every turn.

The full significance of his words on this occasion came up to me only next morning, when I saw him lying on the deck with a murderous weapon in his hand! I was not expecting a cowardly, night attack, nevertheless I kept my gun loaded. I went to sleep this night as usual, forgetting the unpleasant episode as soon as my head touched the pillow; but my wife, with finer instincts, kept awake. It was well for us all that she did so. Near midnight, my wife [Henrietta], who had heard the first footstep on the poop-deck, quietly wakened me, saying, "We must get up, and look out for ourselves! Something is going wrong on deck; the boat tackle has been let go with a great deal of noise, and—O! don't go that way on deck. I heard some one on the cabin steps, and heard whispering in the forward entry."

"You must have been dreaming," I said.

"No, indeed!" said she; "I have not been asleep yet; don't go on deck by the forward companionway; they are waiting there, I am sure, for I heard the creaking of the loose step in the entry."

If my wife has not been dreaming, thought I, there can be no possible doubt of a plot.

Nothing justifies a visit [by a crewman] on the poop-deck after working-hours, except a call to relieve sickness, or for some other emergency, and then secrecy or stealth is non-permissible.

It may be here explained to persons not familiar with ships, that the sailors' quarters are in the forward part of the ship where they (the sailors) are supposed to be found after working-hours, in port, coming never abaft the mainmast; hence the term "before the mast."

My first impulse was to step on deck in the usual way, but the earnest entreaties of my wife awoke me to a danger that should be investigated with caution. Arming myself, therefore, with a stout carbine repeater, with eight ball cartridges in the magazine, I stepped on deck abaft instead of forward, where evidently I had been expected. I stood rubbing my eyes for a moment, inuring them to the intense darkness, when a coarse voice roared down the forward companionway to me to come on deck. "Why don't ye come on deck like a man, and order yer men forid?" was the salute that I got, and was the first that I heard with my own ears, and it was enough. To tell the whole story in a word, I knew that I had to face a mutiny.

I could do no less than say: "Go forward there!"

"Yer there, are ye?" said the spokesman, as with an oath he bounded toward me, cursing as he came.

Again I ordered him forward, saying, "I am armed,—if you come here I will shoot!" But I forbore to do so instantly, thinking to club him to the deck instead, for my carbine was a heavy one. I dealt him a blow as he came near, sufficient, I thought, to fell an ox; but it had, apparently, no effect, and instantly he was inside of my guard. Then grasping me by the throat, he tried to force me over the taffrail, and cried, exultingly, as he felt me give way under his brute strength, "Now, you damn fool, shoot!" at the same time drawing his knife to strike.

I could not speak, or even breathe, but my carbine spoke for me, and the ruffian fell with the knife in his hand which had been raised against me! Resolution had proved more than a match for brute force, for I then knew that not only my own life but also the lives of others depended on me at this moment. Nothing daunted, the rest came on, like hungry wolves. Again I cried, "Go forward!" But thinking, maybe, that my rifle was a single shooter, or that I could not load it so quickly, the order was disregarded.

"What if I don't go forward?" was "Bloody Tommy's" threatening question, adding, as he sprang toward me, "I've got this for you!" but fell instantly as he raised his hand; and there on the deck was ended his misadventure! and like the other he fell with the deadly knife in his hand. I was now all right. The dread of cold steel had left me when I freed myself from the first would-be assassin, and I only wondered how many more would persist in trying to take my life. But recollecting there were only two mutineers left, and that I had still six shots in the magazine of my rifle, and one already in the chamber, I stood ready with the hammer raised, and my finger on the trigger, confident that I would not be put down.

There was no further need of extreme measures, however, for order was now restored, though two of the assailants had skulked away in the dark. How it was that I regained my advantage, after once losing it, I hardly know; but this I am certain of, that being down I was not to be spared. Then desperation took the place of fear, and I felt more than a match for all that could come against me. . . .

It was significant, the court thought afterwards, that while my son had not had time to dress, they all had on their boots except the one who fell last, and he was in his socks, with no boots on. It was he who had waited for me as I have already said, on the cabin steps that I usually passed up and down on, but this time avoided. Circumstantial evidence came up in abundance to make the case perfectly

> clear to the authorities. . . . I will only add that no indignities were offered me by the authorities through all the strict investigation that followed the tragedy.
>
> The trial being for justice and not for my money the case was soon finished.
>
> *Voyage of the "Liberdade"* by Capt. Joshua Slocum, 1890

Would you have responded any differently in those circumstances? If an assailant has you by the throat over the taffrail and is pulling a knife, the time for mediation has probably passed.

Less easy to deal with is the claim that Slocum raped a twelve-year-old girl. Some biographers have tried to dismiss this by suggesting the evidence against him was flimsy. However, similar accusations had been made previously and, in the case of the twelve-year-old, he pleaded "no contest" in court. Beware the perils of hero worship.

On 14 November 1909 Slocum set sail on the *Spray* out of Cape Cod for the Caribbean. He was never seen again; no doubt another victim of the mysterious BERMUDA TRIANGLE.

sloop

a FORE-AND-AFT rigged sailing boat with a single FORESAIL (headsail); except in the US, where a sloop could have two foresails. If a sloop had two foresails in Britain, it would be a CUTTER. In the US, a cutter has a reefable foresail extended to a BOWSPRIT. As Winston Churchill said, "We are two nations divided by a common language."

small-boat sailors

people who sail small boats. These days the mass media is very taken with sailors who race each other or the clock around the world. Even resourceful solo circumnavigators have faded into relative obscurity. If you are not at the glamour end of the scene and are content to potter around "your" estuary, varying it with the occasional mad excursion beyond the headland, worry not—you would have captured Jack London's imagination.

> A sailor is born, not made. And by "sailor" is meant, not the average efficient and hopeless creature who is found to-day in the forecastle of deepwater ships, but the man who will take a fabric compounded of wood and iron and rope and canvas and compel it to obey his will on the surface of the sea. Barring captains and mates of big ships, the

> small-boat sailor is the real sailor. He knows—he must know—how to make the wind carry his craft from one given point to another given point. He must know about tides and rips and eddies, bar and channel markings, and day and night signals; he must be wise in weather-lore; and he must be sympathetically familiar with the peculiar qualities of his boat which differentiate it from every other boat that was ever built and rigged. He must know how to gentle her about, as one instance of a myriad, and to fill her on the other tack without deadening her way or allowing her to fall off too far.
>
> "Small-Boat Sailing" from *The Human Drift* by Jack London, 1911

There you are, "the small-boat sailor is a real sailor." See SKIFF for more Jack London.

snap shackle

see SHACKLE.

snatch block

a BLOCK that can be opened at one side to allow a rope to be dropped into it, rather than be fed through.

snub

to drop the ANCHOR suddenly as a means bringing the vessel to a sudden stop; or, when the sea lifts the bows of a vessel causing it to jerk against the anchor chain, it is said to have "snubbed."

snubber

a temporary line from a cleat or other strong point near the BOW of a boat to the ANCHOR cable; used to share load on both yacht and GROUND TACKLE.

snugged down

sailing with just the right amount of sail for the prevailing wind conditions.

Sod's Law

see MURPHY'S LAW.

soft-tack

see BREAD.

soldier's wind

a wind that comes onto the BEAM of a vessel is a "soldier's wind"; an easy point of sailing that demands little by way of nautical skills.

sole

1. the floor in a vessel; cockpit sole, cabin sole, etc.
2. a particularly tasty flat fish; Dover Sole.

solo sailing

single-handed sailing; an activity in direct contravention of COLREGS (US: NAVRULES) International Rule 5:

> **Look-out**
>
> Every vessel shall at all times maintain a proper look-out by sight and hearing as well as by all available means appropriate in the prevailing circumstances and conditions so as to make a full appraisal of the situation and of the risk of collision.

This rule is breached immediately you go below to get the CHICKEN SUPREME on or to "crash out" in the BERTH, very tired after the first day's run. But, as with many other sailors, I *like* going single-handed. There are a fundamental reason for this: I snore so badly that fellow crew members have been known to jump overboard in order to escape. Either that or they transfer their sleeping bags to the INFLATABLE towed by a handy 200ft WARP. Even then, some have been known to cut the warp. In any case . . .

> If the traveller cannot find
>
> Master or friend to go with him,
>
> Let him travel alone
>
> Rather than with a fool for company.
>
> Buddha, from *Dhammapada*, Chapter 5

sonar

a method of locating underwater objects and determining their range by transmitting sound waves and measuring the time it takes any resulting echo to bounce back. Used in DEPTH FINDERS, DEPTH SOUNDERS.

"son of a gun"

the preferred location for childbirth on naval square-riggers was the gun-deck with its flat, scrubbable boards and rings to hold on to. The mothers were usually women of the waterfront who would provide "rest and recreation" for the sailors onboard ship (the pressed men not being allowed ashore). The fathers were . . . largely unknown. Hence "son of a gun," a bastard. Incidentally, the expression "show a leg" comes from the petty officer's need to check the gender of the occupants of HAMMOCKS when mustering the WATCH. Anything smooth and delicate would be allowed to stay put; anything hairy was required on DECK. I'm uncertain about the reliability of this test.

sounding

the depth of water under the vessel or the process of checking the depth; see LEAD-LINE.

south cardinal buoy

part of the IALA system of BUOYS used to mark the southern limit of a hazard to navigation. Characterized by a TOPMARK of two cones pointing downward, colors yellow over black and a quick-flashing or very quick-flashing white light showing six short followed by one long. See also CARDINAL BUOYS.

sou'wester

1. any wind originating from the southwest.
2. name for a large-brimmed waterproof (OILSKIN) hat once much in favor with fishermen and other seafarers.

spar

a wooden, metal, or synthetic material support for a yacht's rigging; a generic term for BOOMS, YARDS, GAFFS, and POLES.

special marks

a system of BUOYAGE used to define areas reserved for specific purposes such as bathing; also used to define channels for deep-drafted ships; are yellow in color with an X-shaped TOPMARK and yellow light. See IALA.

speed

Sailing boats whose hulls cut *through* the water have a theoretical maximum speed that is based on the length of the hull at the WATERLINE. Once that maximum speed has been reached, any additional sail or engine power brought into play will be wasted. This does not apply to craft such as racing dinghies like the Fireball or catamarans such as the Hobie-cat. These have hulls that are designed to aquaplane and, once the hulls are out of the water, the rule does not apply. The formula is:

$s = 1.4 \times \sqrt{LWL}$		
where:		
s	=	speed in knots
$\sqrt{}$	=	square root
LWL	=	length on the waterline in feet

So, for a range of typical boat lengths, the maximum speeds are:

LWL		Speed
26ft		7.1 knots
32ft		8 knots
37ft		8.5 knots
42ft		9 knots
47ft		9.6 knots

So no one's going anywhere fast, really. (To perform the above calculations using metric, convert the LWL from feet to meters and then apply the formula.)

spill, to

to release wind from a sail by easing the sheet or LUFFing to WINDWARD; slows the boat down.

spinnaker

a very full three-cornered sail related to the parachute. Flown from the MASTHEAD beyond the FORESAIL, can add considerable speed to downwind RUNNING. Its manageability problems make it one of the least-used sails cluttering up the forecabin.

spitfire jib

the last resort; a tiny FORESAIL made from very heavy sailcloth or three-ply timber used in extreme weather conditions.

splice, to

to join two ropes together by un-LAYING the ends and over-LAYING them back together again.

sponge

vernacular expression used to refer to *poriferan*, a type of aquatic organism and the preferred alternative to the horrible pink or yellow things sold in chain store pharmacies.

Recommended source: try the island of Simi (or Symi) in the Dodecanese group of Greek Islands. The shipwrights of Simi were once famed for building the fast galleys used by the Knights of St. John. Simi is not a package holiday destination for North Europeans and most visitors come on the ferries from nearby Rhodes. So, if not exactly the Greek Islands at their best, the place is relatively unspoiled. Visiting yacht-borne sponge hunters should moor stern-to on the north side of Simi Town Harbor, as far to the west of the ferries and the handsome clock tower as possible. Once ashore, head west for the best tavernas. Head east toward the clock tower looking for a sponge shop named (in English) "Famous Bob's Sponges." Now Bob—Bob Taylor—is not really all that famous, but he used to be a journalist who worked for British newspapers such as the *Yorkshire Post* and the tabloid *News of the World*. Bob packed it in early after a couple of heart attacks (God's way of frowning on his profession, I told him) and he took the smart option of retiring early to Simi. Now he will give you the low-down on sponges and sell you the best at a very reasonable price.

spoon

to spoon; see SCUD.

spreaders (US)

See CROSSTREES.

springs

lines used when mooring a vessel to inhibit its to and fro movement; springs run from a CLEAT toward the STERN of the boat to a BOLLARD forward of the BOW (and vice versa). Springs are also used when maneuvering a yacht in harbor.

spring tides

high and low tides furthest from the mean (i.e., with the greatest RANGE); caused by the sun and moon being in conjunction (high) or opposition (low). Probably the only instance where there is any veracity in astrology.

sprit, spritsail

a sprit is a spar that diagonally crosses a square-shaped fore-and-aft sail, usually from a "shotter" at the foot of the main mast up to the outer-most EARING of the sail. The bowsprit seems to be the exception that proves the rule. The best-known spritsail rig is the Thames Barge (which also has a bowsprit).

squadron

a group of warships detached on a particular duty or under the command of a flag officer. In the (British) Royal Navy it can refer to a subdivision of a larger fleet. In the US Navy it usually refers to a group of destroyers and submarines.

(More here on the adventures of Commander John Byron. See also GIANTS OF PATAGONIA, MARINES, MIDSHIPMAN, and MUTINY.)

As soon as Byron returned from his miserable trip to Patagonia on HMS *Wager* (see MUTINY), he was awarded his own ship as post captain of HMS *Siren*. Still only twenty-three, his foot was now firmly on the promotion ladder.

Sure enough, in 1764 he was awarded command of a new squadron with a top-secret mission. The Admiralty sent *Commander* Byron off in the COPPER-BOTTOMED HMS *Dolphin* with the ambitious orders to seize for Britain the Falkland Islands, California (New Albion) and to explore the Northwest Passage linking the Pacific and Atlantic Oceans.

The little squadron consisted of his flagship, the FRIGATE *Dolphin*; the SLOOP *Tamar*; and a storeship called the *Florida*. (Judging by her name, HMS *Florida* was probably a prize won from the Spanish navy.)

In the 18th century frigates were three-masted, squared-rigged ships that compromised the heavy fire-power of the "ships of the line" in favor of speed and agility. Nelson's flagship HMS *Victory* sported 104 heavy cannon on three decks; thirty of these were 32-pounders (that's the weight of the cannonballs, not the guns of course). The *Dolphin* made do with twenty-four guns on a single deck and these were pretty lightweight 9-pounders. Her length overall was 34 meters and 10 meters in the beam; she was crewed by 160 sailors. HMS *Tamar* was a smaller three-master with just 16 guns on a single deck. The *Florida* may have had three-pounder anti-personnel swivel cannons on her QUARTERDECK and FO'CASTLE but would have relied on *Dolphin* and *Tamar* for protection. Her main task was to keep up with the others.

In 1764 when Byron faced their lordships in the OLD ADMIRALTY building he must have experienced conflicting emotions. He would have been puffed up with pride at being given command of an important expeditionary squadron. His previous squadron command four years earlier had required him to go to Louisbourg in Nova Scotia. ("New Scotland." The French were calling Quebec "New France." It was a battle of names.)

Byron's mission was to destroy a fort that had been won from the French two years earlier and needed to be permanently denied to them. While he was there a French flotilla turned up to replenish settlements further up the Lawrence River in Quebec. They were engaged by Byron and heavily defeated, putting an end to France's colonial ambitions in North America.

But the Hon. John Byron had nagging doubts about his new tasking. The relatively fine weather of the west coast of North America would have appealed. It was 185 years since Sir Francis Drake had walked up a beach and made a clear signal of intent to the Spanish by calling all he surveyed "Nova Albion." ("New Britain" in Latin; but it probably didn't fool anyone.) Returning there to plant the Union Flag would be a fine achievement for Byron. But to get there his squadron would have to sail around the dreaded coast of Patagonia; he'd spent enough character-forming years there when HMS *Wager* was wrecked.

He must have considered this: if he successfully seized the Falklands against Spanish and French opposition, then made it around Cape Horn into the Pacific for a great stopover in California, what faced them next? Up through the Bering Sea to the frozen north and another futile attempt to find a passage through the Arctic pack ice to the North Atlantic? Europe's 18th-century super-powers Britain, France, and Spain were desperate to establish a speedy trade route to the riches of the Orient. A lot of ships and lives had already been lost in trying to open the Northwest Passage. It must have been a chilling prospect even for an experienced seaman like Byron. *Especially* for a seasoned seaman like Byron. (Such a shortcut did open up when the polar ice started to melt, but that wasn't until 2009.)

The admirals must have anticipated his concern because they had a cunning plan. Commander Byron would not tell his crews where they were headed until they were too far south in the Atlantic to turn back. Then he would address them from the quarterdeck, look at their miserable faces, and announce that they were all on double pay!

The idea worked. But what Byron hadn't told the admirals back home was that he'd absolutely no intention of heading for the frozen north on what he saw as a suicide mission.

From his previous circumnavigation Byron knew that the Solomon Islands were a much-preferred destination after leaving California. He'd discover many new islands. Maybe he'd be the first to claim Terra Australis Incognita ("Unknown Land Down-Under") for Britain? If he did that the Admiralty would forgive him for passing on the Northwest Passage.

In June 1764 he set sail from England and headed south. The Seven-Year' War with France was over (Britain won) but he no doubt kept a weather eye open for French and Spanish PRIVATEERS. They made good progress down to the west coast of Africa but when they struck out for Brazil, the three ships got caught in the DOLDRUMS. This tropical region of calms between the northern and southern trade winds is a nightmare for sailors. There is no wind to ease the blistering heat, no wind to fill the sails and make headway. Boredom is only punctuated by the occasional thunderstorm.

It took weeks for the squadron to reach Rio de Janeiro, the preferred port-of-call for victualing and repairs; they arrived on 12 September 1764. For the crew there would be rest and recreation.

However, John Byron was aware of what Rio de Janeiro had done to the naval career of William Dampier. Dampier (1651–1715) was an explorer and privateer, a triple-circumnavigator who is more feted as the man who stranded Alexander Selkirk on Juan Fernández Island off the coast of Chile as punishment for complaining that their ship, HMS *Cinque Ports*, was unseaworthy. It was the incident that inspired Daniel Defoe's *Robinson Crusoe*. Selkirk was right: a month later the *Cinque Ports* sank. William Dampier was a considerable sailor but he did not understand the law of unintended consequences. On his previous voyage he'd put another sailor ashore—this time in Rio. That doesn't sound too bad but the Brazilian police jailed George Fisher. When Fisher got home he complained about being banged up abroad. When Dampier got home the Admiralty found him guilty of cruelty, docked all his pay for the voyage and threw him out of the navy.

Like Dampier, Byron had bother with the Brazilians. When he gave his crew shore leave in Rio, nine men failed to return. Freelance press-gangs were a problem in the port. Local gang-masters would grab sailors off the streets and keep them under lock and key until their ships had set sail. Then they'd be supplied to the next vessel in need of crew—for a commission. An interesting business model. But witnesses told Byron that his sailors had disappeared from the dockside taverns arm-in-arm with exotic young ladies. The innovative idea of a female press-gang was an even more interesting business model.

Byron was having none of it: he wanted his men back. He sent other sailors to ask questions around the bars, maybe spread a little gold coin around, and learned where the crew members were being held hostage. Then he mustered a raiding party and rescued them. Whether the seamen were grateful or not isn't recorded.

The first big man John Byron had to deal with in South America was Major-General Robert Clive, the soldier who, through the vehicle of the East India Company, had won India for the British. Known in newspaper headlines as "Clive of India" he was the son of a land-owning lawyer and local member of Parliament for the Shropshire town of Market Drayton. As a lad Clive junior had been expelled from school three times and terrified the locals by climbing the tower of St. Mary's Parish Church and sitting on a gargoyle.

Unhappy with his personal allowance Clive came up with his own business plan. He would tour the town telling shopkeepers that, in return for a small weekly stipend, he could guarantee they had no trouble with local scoundrels and vagabonds. Remarkably, the hard-working retailers who turned him down would be woken in the night by the sound of breaking glass; the vagabonds were on Clive's payroll. If you think this sounds like what we now call a protection racket, you'd be right. Clive of India was a man who liked to get his own way and was just the chap to plunder the riches of India for King George III.

John Byron was living in a comfortable rented house while Portuguese shipwrights busied themselves CAULKING his ships. He reported that he and the crews were enjoying "as much fresh meat and greens as they could eat every day" and were keeping the scurvy at bay. This is what he wrote about the arrival of Maj. Gen. Clive.

> While we lay there, Lord Clive, in the *Kent Indiaman*, came to the port. The ship had sailed from England a month before us, and had not touched anywhere, yet she came in a month after us; so that her passage was two months longer than ours.
>
> *An Account of a Voyage Around the World in the Years MDCCLXIV, MDCCLXV, and MDCCLXVI by the Honourable Commodore Byron in His Majesty's Ship the Dolphin*, 1767

Clive's ship saluted the *Dolphin* with an eleven-gun salute, a courtesy that was returned in kind. The *Kent Indiaman* was in an appalling condition and repairs were not going to be started until the best local workmen had finished working on Byron's ships. Desperate to reach Bengal, Clive demanded that Byron take him onboard the *Dolphin* and set sail for the Indian Ocean. Byron, totally outranked but probably offended at being treated like a taxi driver, told him he was on a secret mission for the Admiralty and his orders could not be varied.

Byron's squadron left Rio de Janeiro on 9 October and headed south toward the Falkland Islands. But there was one more item on his to-do list, a task that required him to go ashore in Patagonia. See GIANTS OF PATAGONIA.

squall

a sudden and very strong gust of wind (sometimes with a shower of heavy rain) lurking in the aftermath of a depression for the unaware.

"square meal"

a "square meal" is thought to refer to a square of wood provided by the ship's carpenter to seamen who could not provide their own plates; more likely to mean a fair ration ("square" in the sense of "honest").

square rig

the principal sails at right angles to the length of the ship and extended by horizontal YARDS slung by the middle.

stanchion

the posts the hold up the bulwarks or LIFELINES.

standard port (UK)

equivalent to a REFERENCE STATION in US tide tables; high and low tides (times and heights) are published for each day for every standard port throughout the year, adjustments being made where necessary for intermediate secondary ports.

stand by

1. an order to be ready, to be prepared.
2. a phrase used in radio-telephony; "stand by," wait for me to call you back soon or "standing by," waiting for you to call me back soon.

standing part

the main, non-working component of a rope, wire, or line.

standing rigging

see RIGGING.

stand off

to steer or stop clear of something is to "stand off"; probably the origin of the expression "stand offish," meaning distant, antisocial.

stand on

to steer closer to or to maintain a course; see STAND-ON VESSEL.

stand-on vessel (action by)

the stand-on vessel in a right-of-way situation is one that need not take avoiding action; it does however have responsibilities as described in ColRegs International Rule 17:

Action by Stand-on Vessel

(a)
- (i) Where one of two vessels is to keep out of the way the other shall keep her course and speed.
- (ii) The latter vessel may however take action to avoid collision by her manoeuvre alone, as soon as it becomes apparent to her that the vessel required to keep out of the way is not taking appropriate action in compliance with these Rules.

(b) When, from any cause, the vessel required to keep her course and speed finds herself so close that collision cannot be avoided by the action of the give-way vessel alone, she shall take such action as will best aid to avoid collision.

(c) A power-driven vessel which takes action in a crossing situation in accordance with subparagraph (a)(ii) of this Rule to avoid collision with another power-driven vessel shall, if the circumstances of the case admit, not alter course to port for a vessel on her own port side.

(d) This Rule does not relieve the give-way vessel of her obligation to keep out of the way.

starboard

the right-hand side of a vessel when facing forward; the side the "steering board" used to be fitted in the days before the RUDDER took over as the preferred method of getting a boat to point in the right direction. See LARBOARD.

starboard tack

sailing CLOSE-HAULED with the wind coming over the STARBOARD side of the boat; confers right of way when two boats under sail and on different tacks are on converging courses.

star chart

a map of the night sky that assists in the identification of FIXED STARS for navigation purposes; you might have two, one for the Northern Hemisphere, another for the Southern Hemisphere. A star chart will not show the sun, moon, planets, or comets because the position of these is too variable.

stars

see CELESTIAL OBJECTS and FIXED STARS.

stateroom

salesperson-speak for a big CABIN.

statute mile

distance of 5,280ft; standard measure of distance on land in the United States and, surprisingly, Britain. Also used, note, on inland waterways in those countries; elsewhere, kilometers. Not to be confused with the *nautical* mile.

stay

a line (usually made of heavy wire) used to support a mast in the FORE-AND-AFT direction; see FORESTAY, BACKSTAY.

staysail

a triangular sail on a cutter located between the JIB and the MAIN MAST and hanked onto a fractional FORESTAY; or a small, stiff sail rigged between the main boom and mast.

steaming light

a white light fitted to the head of the mainmast or near the cross-trees on a sailboat to indicate that the vessel is using auxiliary power.

steerage way

moving fast enough through the water for the RUDDER to be effective.

stellar magnitude

see MAGNITUDE OF STARS.

stem

see BOW.

step

socket or box that holds the foot of a MAST.

stern

the back of a vessel; the AFT.

stern board

when a yacht is in IRONS, the wind will eventually push her astern fast enough for the RUDDER to have effect. She is then making "stern board" and can be steered around onto one TACK or the other.

stern drive

power and transmission arrangement comprising an INBOARD ENGINE connected to a steerable drive system extending through a cut-out in the TRANSOM to the PROPELLER.

sternpost

a vertical post at the TRANSOM to which the RUDDER is attached.

stevedore (US)

literally, a packer of cargo. The word came into use in American English via the Spanish word *estivador*.

stiff

a stiff yacht is one that does not easily HEEL (lean with the wind). I like stiff yachts—they are easier to cook in.

stingrays

this fish can be particularly dangerous because it swims into sandy shallows and beaches where you may be wading.

In 1863, young Louis De Rougemont found himself in two survival situations at the same time. While on the run from his Swiss school (to dodge French military conscription), he bought himself a berth on a Dutch pearl-fishing boat. The boat came to grief and Louis was soon making do on an island in what we now call Indonesia.

> After breakfast, my dog and I would go out to catch a peculiar sort of fish called the "sting-rae." These curious creatures have a sharp bony spike about two inches in length near the tail and this I found admirably adapted for arrowheads. The body of the fish resembled a huge flounder, but the tail was long and tapering. They would come close in-shore, and I would spear them from the rocks with a Papuan fishing-spear. The smallest I ever caught weighed fifteen pounds, and I could never carry home more than a couple of average weight. They have the power of stinging, I believe, electrically, hence their name. At all events, I was once stung by one of these fish, and it was an experience I shall never forget. It fortunately happened at a time when some friendly blacks were at hand, otherwise I question very much whether I should be alive to-day.
>
> I was wading slowly along the beach in rather deep water, when I suddenly felt a most excruciating pain in my left ankle. It seemed as though I had just received a paralysing shock from a powerful battery, and down I fell in a state of absolute collapse, unable to stir a finger to save myself, although I knew I was rapidly drowning. Fortunately the blacks who were with me came and pulled me ashore, where I slowly recovered. There was only a slight scratch on my ankle, but for a long time my whole body was racked with pain, and when the natives got to know of the symptoms they told me that I had been attacked by a "sting-rae." The spike or sting measures from two to six inches in length according to the size of the fish.
>
> *The Adventures of Louis De Rougemont*, by Himself, 1898?

Louis was lucky to have some local expertise at hand. The tail of the stingray carries a venom (not a high voltage capacitor) that can be fatal.

In 2006 Australian conservationist and wildlife expert Steve Irwin was killed when a stingray's barb pierced him in the chest while he was filming on the Great Barrier Reef.

See HAZARDOUS MARINE CREATURES.

stock

the crossbar of a fisherman's ANCHOR, being fixed to the SHANK at the opposite end from and at right angles to the FLUKES; serves to ensure that the fluke is best able to hook onto or dig into the sea bed.

stop

1. to stop: to prevent the end of a rope from fraying or coming apart by WHIPPING or taping.
2. something a boat does only when it is out of the water.

stopper knot

a knot fashioned in the end of a rope to prevent it from running out through block or eye; the figure-of-eight knot is the preferred choice.

stores

food provisions stowed onboard a yacht.

storm

see BEAUFORT SCALE.

Storm Tactics

subtitled "Modern Methods of Heaving-To for Survival in Extreme Conditions," *Storm Tactics Handbook* by Lin and Larry Pardey is a superb book. It should be studied and its advice implemented by all yachtsmen and yachtswomen wherever they sail—storms do not limit themselves to the high seas.

stoves (cookers)

a means of cooking food and incorporating two, three, or four burners (hobs), a grill (US: broiler) and, in most cases, an oven. On a boat you have all the options available on shore with the exception of electricity (microwave ovens are described below).

For centuries, solid fuel cookers burning coal or wood were the warm heart of GALLEYS on boats and ships. The skippers of today's vintage sailing boats often use modern purpose-built stoves that provide heat as well as a single hot plate on top for cooking. But for a sailor going beyond coastal cruising, a reliable supply of coal or wood is a big disincentive.

A more practical approach was adopted with the introduction of stoves that burned paraffin oil. The upside of these is that they are considered to be safe on boats and the fuel is relatively cheap. Availability is a problem though; if you are berthed-up in Florida, for example, it can be quicker to sail to the Bahamas for a new supply than to try and get paraffin in the local stores. In "the Islands" it is used for lighting and heating because local electricity supplies are unreliable. This is not true in Florida.

Getting a paraffin stove started is another problem because the burners will not work until they have been preheated. This is done by pouring medicinal alcohol into a little tray under the burner. This ignites easily. Meanwhile, the pressure chamber is pumped up, causing the paraffin to vaporize. Turn the valve to release the vapor and boom, you're cooking.

The third downside is that paraffin stoves produce more smoke than heat and your shiny bright saloon soon resembles the working end of a coal mine, only not smelling as nice. This goes a long way toward explaining the switch to liquid petroleum gas (LPG) products, specifically propane and butane. (Butane gas is the sort of stuff that comes in cans and is used by campers.)

For safety reasons care is needed in the planning of a propane gas system on a boat. Propane is heavier than air and if any part of the system leaks, the gas will descend to the bilges. A naked flame of a spark could then cause the boat to explode. For this reason, propane bottles need to be stowed externally in a special compartment that drains off over the side. An alarm system connected to a "sniffer" in the bilges is also recommended. This will sound an alarm and automatically shut off the gas supply. Good practice suggests that gas should be burned out of the supply pipes after each cooking session.

stow

to store anything onboard a vessel.

stowage

any location on a boat designated as a place where things are stored—or stowed.

stowaways, ghosts, and alien abductions

I am not the sort of person who believes in ghosts. If you brought a ghost to dinner, I still would not believe in them. However, some solo sailors after many days at sea and often in severe weather begin to hallucinate. Even the redoubtable Captain Joshua SLOCUM experienced this. *Spray* was sailing out of Horta in the Azores heading for the Canary Islands when trouble struck.

> I am a careful man at sea, but this night, in the coming storm, I swayed up my sails, which, reefed though they were, were still too much in heavy weather; and I saw to it that the sheets were securely belayed. In a word, I should have laid to, but did not. I gave her the double-reefed mainsail and the whole jib instead, and set her on her course. Then I went below, and threw myself upon the cabin floor in great pain.
>
> How long I lay there I could not tell, for I became delirious. When I came to, as I thought, from my swoon, I realised that the sloop was plunging into a heavy sea, and looking out of the companionway, to my amazement, I saw a tall man at the helm. His rigid hand, grasping the spokes of the wheel, held them as in a vise. One may imagine my astonishment. His rig was that of a foreign sailor, and the large red cap he wore was cockbilled over his left ear, and all was set off with shaggy black whiskers. He would have been taken for a pirate in any part of the world.
>
> While I gazed at his threatening aspect, I forgot the storm, and wondered if he had come to cut my throat. This he seemed to divine.
>
> "Señor," he said, doffing his cap, "I have come to do you no harm." And a smile, the faintest in the world, but still a smile, played on his face, which seemed not unkind when he spoke. "I have come to do you no harm. I have sailed free," he said, "but was never worse than a *contrabandista*. I am one of Columbus's crew," he continued. "I am the pilot of the *Pinta* come to aid you. Lie quiet, señor captain," he added, "and I will guide your ship tonight. You have a *calentura*, but you will be all right tomorrow."
>
> I thought what a very devil he was to carry sail. Again, as if he read my mind, he exclaimed:
>
> "Yonder is the *Pinta* ahead; we must overtake her. Give her sail; give her sail! *Vale, vale, muy vale!*" Biting off a large quid of black twist, he said: "You did wrong, captain, to mix cheese with plums. White cheese is never safe unless you know whence it comes. *Quien sabe*, it may have been from *leche de Capra* and becoming capricious—"
>
> "Avast, there!" I cried. "I have no mind for moralizing."

So, the truth is out! This *contrabandista* knows his stuff. What was that about cheese? It all started, it seems, back in the Azores.

> Since reaching the islands I had lived most luxuriously on fresh bread, butter, vegetables, and fruits of all kinds. Plums seemed to be most plentiful on the *Spray*, and these I ate without stint. I also had a Pico white cheese that General Manning, the American consul-general, had given me, which I supposed was to be eaten, and of this I partook with the plums. Alas! By night-time I was doubled up with cramps.

Alas! indeed. The weird visions seem to be down to the cheese.

Edward Lear was certainly on "no end of Stilton Cheese" when he wrote the *JUMBLIES*. The evidence is now all stacking up to present a convincing solution the baffling question of alien abductions. The answer lies in the predilection for junk food—it is caused by all that cheese in burgers and pizzas. See also SLOCUM.

strand

1. the lines or wires from which a rope or cable is woven.
2. a beach.

strum box

If you have bilge, you need a BILGE PUMP. If you have a bilge pump, you need a strum box. The sucking end of the pipe to the pump goes into the strum box and the strum box filters out all the grunge slopping around in the bilge, thus preventing it from blocking the pump.

strut

a fitting under the STERN of a hull used to support the exterior portion of a propeller shaft.

stuffing box

Of all the holes we have to make through our "watertight" hull, the one through which the propeller shaft passes is the one most likely to cause problems. The stuffing box is a fitting that supports the shaft and lubricates it. It does of course have to be watertight, too.

subordinate station (US)

equivalent to a SECONDARY PORT in UK tide tables; an intermediate point between REFERENCE STATIONS for which an additional calculation has to be made. See TIDE TABLES.

sump

the "sump" or "shower sump" is the lowest part of the BILGES and the location for an automatic pump that will eject the water out via a THROUGH-HULL valve.

"sun is over the yardarm"

past midday; time for the first drink of the day.

survival suits (dry suits)

Any crew member going into the water, for whatever reason, is far more likely to die from exposure than from drowning. For this reason alone I would consider survival suits to be a good buy for anyone likely to find themselves sailing in extreme weather conditions. A survival suit is totally waterproof with elasticated wrists, ankles, and neck keeping the water out and your body heat in. They are not cheap, typically being priced at £200 ($300).

Tip: In the UK at least it is possible to buy survival suits previously used by oil companies flying workers by helicopter to and from North Sea oil platforms. These can be available for as little as £50 ($75) but, obviously, examine them very closely before buying.

"swallow the anchor"

to retire from the sea to live ashore; or not, as the case may be.

"JB" and I became good friends during a few months stopover on an island of character. We were both at anchor in a bay on the east side when the weather backed round and threatened to leave us exposed to a lee shore. JB's family had tried to stop him going sailing for the not-outrageous reason that he was seventy-eight years old. But JB had been born and raised in a New England lighthouse (honestly!), spent the 1930s in commercial fishing, spent World War II on North Atlantic Convoy patrols, and then became a professional yacht skipper. His last job before retiring was to supervise the building of a $25M super-yacht in Italy; he'd had enough of the lifestyle. JB's compromise with his family was to agree not to sail single-handed. He'd reached the island with a younger buddy who was a retired firefighter. While we were getting ready to up-anchor and move round to the lee of the island, we learned that the firefighter was "on-call" with a very friendly French nurse; JB could only make the trip if he sailed solo. We had two boats and only three crew. Fortunately, the firefighter and the nurse turned up at the last moment and all was well.

A couple of weeks later, JB invited me around to his boat for supper. This was my reward for fixing his shortwave weather fax. The old guy was the most impressive ocean voyager I had ever met and he was a great storyteller. But, surprisingly, he was no cook. I seem to have driven the meal itself deep down into my subconscious mind; what I do remember is that he served it on paper plates, the kind you might get at a party or a barbecue.

JB will never "swallow the anchor."

swamp

to fill a boat with water and potentially disastrous consequences; see BAIL.

swell

an undulating, non-breaking surface of the sea; once the wind has dropped or the WAVES have moved away from the effect of the "influencing force," the waves are said to "mature" into a regular pattern known as a "swell." In maturing, waves resolve themselves into groups with similar wavelengths and period (speed) (see WAVES for definitions of these terms). Groups of waves with the same origin and direction are known as "wave trains." Unimpeded, waves in the form of "swell" can travel for many hundreds of miles in the same direction. It was this settled, undulating form of wave that was most useful to ancient navigators without magnetic compasses; its consistency would help them hold a course. The origin of a swell can be somewhat random: "five days ago there was a storm somewhere to the south." But the origin can also be predictable to a degree; the northeast trade winds of the North Pacific are known to generate a northeast swell. Cross-swells caused by conflicting disturbing forces or by swells reflected off land masses can make this form of steering very difficult, as do waves caused by local surface winds working across the direction of the underlying swell. In such confused seas, and in the absence of any other clues to direction, a helmsman can do no better than struggle to keep any kind of decent course.

swing

an anchored or moored vessel swings when the wind and/or tide changes her position in relation to the mooring.

synoptic chart

a large area weather map showing the location of depressions, high-pressure areas, and fronts using isobars to indicate lines of equal pressure; an essential tool for onboard weather forecasting and now available via short-wave fax and satellite communications.

T

tabernacle

a wooden or metal hinge at the foot of a mast that enables the mast to be folded down along the DECK; used to enable smaller cruising boats to be sailed on inland waters where they might have to pass under low bridges or to be trailed cross-country on public roads.

tack

1. the foremost lower corner of a FORE-AND-AFT sail.
2. a change in course with the purpose of sailing with the wind across the opposite side of the boat; see GOING ABOUT; also a noun, "Which tack are you on?" to which the answer is PORT or STARBOARD.
3. the kind of thing sold in harbor-side tourist shops and some chandlers'. Better recognized than defined: a hat with "Captain" on it (or "First Mate" or "Galley Slave" for the wife); a sweater with an anchor on the chest; a brass-framed print of a SQUARE-RIGGER for the saloon wall; trendy-looking deck shoes that fall apart on contact with salt water; a folding sextant that fits into your pocket but is unable to help you determine which ocean you are on; an ashtray shaped like a ship's wheel; and tie-back lace curtains for your port lights.

tacking

the process of GOING ABOUT from one TACK to another; to sail to WINDWARD in a zigzag fashion.

tackle

an arrangement of ropes and PULLEYS (mounted in "blocks") used to give additional pulling power; from the Low German word *takel*, ship's rigging.

See BLOCK AND TACKLE for fuller explanation of the mechanics.

tackline

a line of about two meters (6ft) in length used to display SIGNAL FLAGS.

taffrail

a rail across the TRANSOM of the vessel; normally used to refer to the PUSHPIT on a yacht.

taffrail log

a device fitted to the taffrail and used to determine speed and distance traveled; see LOG for fuller description.

take in

to HAND a sail; to pull down a SAIL.

take off

a tide takes off when it changes from SPRINGS to NEAPS.

take up

the effect dry wood has of expanding when it comes into contact with water.

Ever had that annoying problem with a wooden door? In the summer it fits perfectly, but come the fall and the rain it swells and won't open. When you get it open, it won't close again. So you take the damn thing off its hinges and plane down the edges. Next summer you get a draft. Wooden boats are like that. You take it out of the water and it dries, opening the gaps between the hull planking. When you put it back into the water it leaks like hell until it has "taken up" (i.e., the planks have swelled again until the boat is TIGHT).

taking the ground

getting stuck; a serious embarrassment all round, as explained by Joseph Conrad.

> "Taking the ground" is the professional expression for a ship that is stranded in gentle circumstances. But the feeling is more as if the ground had taken hold of her. It is for those on her deck a surprising sensation. It is as if your feet had been caught in an imponderable snare; you feel the balance of your body threatened, and the steady poise of your mind is destroyed at once. This sensation lasts only a second, for even while you stagger something seems to turn over in your head, bringing uppermost the mental exclamation, full of astonishment and dismay, "By Jove! She's on the ground!"
>
> And that is very terrible. After all, the only mission of a seaman's calling is to keep ships' keels off the ground. Thus the moment of her stranding takes away from him every excuse for his continued existence. To keep ships afloat is his business; it is his trust; it is the effective formula of the bottom of all these vague impulses, dreams, and illusions that go to the making up of a boy's vocation. The grip of the land upon the keel of your ship, even if nothing worse comes of it than the wear and tear of tackle and the loss of time, remains in a seaman's memory an indelibly fixed taste of disaster.
>
> "Stranded" within the meaning of this paper stands for a more or less excusable mistake. A ship may be "driven ashore" by stress of weather. It is a catastrophe, a defeat. To be "run ashore" has the littleness, poignancy, and bitterness of human error.
>
> *The Mirror of the Sea* by Joseph Conrad, 1903

tall ship

a sailing ship with a tall MAST or masts—usually a term for a full-rigged clipper ship.

tang

any metal fitting on a spar or mast to which the standing RIGGING is attached.

tar

See JACK, JACKTAR.

tarpaulin

a waterproof CANVAS.

taut

pulled tight.

telltale

lengths of wool or strips of cloth secured to each side of a sail ABAFT the LUFF and used as a means of indicating the effectiveness of the airflow.

tempest

jolly bad weather. "Tempest" is one of many words in European languages that originate from the Latin *tempus*. Oddly though, *tempus* means time, as in "tempus fugit"—time flies. The French word for time is *temps*, which also means weather. The same applies to the familiar Italian word *tempo*. Exactly why and how the word for time came to correspond to the word for weather, I have no idea. Maybe it came from the old nautical saying I just invented: "Time flies when you're having good weather"? See also TIDE.

tender

1. a tender boat is one that HEELS easily, one that is not STIFF.
2. any small boat used to service a larger one; in the case of a yacht, this is often an INFLATABLE DINGHY.

tether (US)

a length or rope or webbing used to link a personal safety harness to a secure point on the boat.

thief-proofing

the process of trying to protect your boat from crooks.

One of the main targets for thieves is the inflatable dinghy when you go ashore for a drink or to shop; and it is not something that exactly lends itself to being chained up like a bike. The solution might lie in making it a less attractive item. Start by painting your boat's name along the side in the biggest letters that will fit. Then buy an inflatable repair kit—ideally, one in which the patches are a different color from the main body of the dinghy so you put orange on grey, black on orange, or whatever. Then stick patches at odd angles along the seams to make it appear that the inflatable has all the boating characteristics of a sieve. Obviously, the newer the dinghy, the more important it is to do this. Once back onboard and deflated, the thing will not look too scruffy.

If the bad people do not like the look of your inflatable, they might decide to retaliate and have off with the outboard or even the oars. If you don't fancy walking down the pier with the outboard on your shoulder (like in those old Seagull advertisements), you could try chaining it to the TRANSOM. And if you have an ANCHOR in the dinghy, chain it to that, too. (Don't bother removing parts from the outboard itself—they are too easily replaceable.) I know people who quickly lash the oars to a KEDGE and use the anchor to sink them out of sight. Another method I have seen used to protect hard-hulled tenders is to drill a hole in the bottom, then fit a bung. Once ashore, take the bung with you. This does not, of course, work when you are tied alongside a pier. Also, I would worry about getting back from the bar and forgetting to replace the bung before slipping the boat back in the water. . . .

The approach I tend to use is to look along the pier or beach when you arrive and pick out the character who you think is most likely to steal the inflatable (or parts thereof). Say hello to him (and it will be a "him" in most places) and invite him for a short walk down the beach. Once away from his mates, appoint him as your head of security for your stay in that location. Pay him an appropriate sum "on signature" and promise a small bonus each time you return and find the dinghy still *in situ*. Tell him where you will be and that you want to know if anyone goes anywhere near the yacht. If all goes to plan, you will have someone who will help you with provisions, run errands, and scrub the bottom.

thimble

round or oval metal or plastic inner forming the core of a SPLICED eye of rope or wire.

"three sheets to the wind"

drunk, out of his skull, incapacitated as a consequence of over indulgence of beverages of an alcoholic nature, has origins of a nautical nature, probably to do with the odd idea that a sail could have three SHEETS or more likely that a square-rigged sail must be out of control if three of its four sheets are "to the wind."

throat

the point at which two parts of a rope or wire come together after forming an eye.

through-hull

a fitting or object that goes through the hull; might be the TRANSDUCER for the DEPTH FINDER but, if intended for input of sea water or the output of fluids will incorporate a one-way valve, ideally the right way around.

thunder box

slang for any rudimentary toilet and, therefore, appropriate for anything found on a small boat; see HEADS.

thunderbox

early restroom (US) or loo (UK).

WARNING: the following functional specification is not for the faint-hearted. You may wish to avert your eyes until THWART. The thunderbox was slung over the side or transom. In the bottom of the box was a hole of appropriate diameter. The user would climb into the box, squat over the hole and perform the required functions while enjoying the view (and checking the ropes securing the box) out of the open top. I have never seen a thunderbox on a modern yacht (though holding onto the pushpit rails and sticking your bum over the side certainly is not unheard of). Anyone thinking of using a thunderbox in a bit of a sea should keep in mind that the benefit of a free douche would be countered by the danger of getting your own back.

thwart (pron. "thort")

a seat in a dinghy.

tide

the rhythmical rise and fall—and consequent flow, or CURRENT—of water caused by the gravitational pull of the moon and the sun. Predicting the depth of water at a specific location and time or the time at which the depth will reach a certain level (rising or falling) is one of the fine arts of navigation. Interestingly, "tide" originates from the Norse word for time, thus binding the all-important relationship. See also TEMPEST and TIDY.

tidefall

a crucial (estimated) measurement when deciding where to anchor. If you have anchored for the night with 1m of water under your keel, 1m to 2m tidefall in the next three hours means you are in the wrong place.

tide tables

an almanac, or annual day-by-day listing of high and low tides (usually but not always two of each) for important ports in an area; these are called STANDARD PORTS (UK) and intermediate locations are called SECONDARY PORTS (UK). Full listings of the intermediate state of the tide are included under the former heading, but a calculated adjustment has to be made for secondaries. In the United States the terms are reference stations and subordinate stations respectively. Tide tables need to be mastered by the good navigator if he or she is going to be able to answer questions such as "Will we get over the BAR?" or "If we anchor here, will we go aground in a few hours?"

tidy

An adjective that supposedly draws an analogy with the predictability of tides, but I have never considered tides to be predictable to quite that degree of accuracy (especially those affected by the wind).

tight

a tight boat or ship is one that does not leak; term that apparently dates only from 1971 but also has the older meaning of "watertight."

tiller

a length wood attached to the top of a RUDDER POST, thus enabling the rudder to be turned and the boat to be steered.

> Scorning a turnstile wheel at [*Pequod's*] reverend helm, [she] sported there a tiller; and that tiller was in one mass, curiously carved from the long narrow lower jaw of her hereditary foe. The helmsman who steered by that tiller in a tempest, felt like the Tartar, when he holds back his fiery steed by clutching its jaw. A noble craft, but somehow a most melancholy! All noble things are touched with that.
>
> *Moby Dick; or The Whale* by Herman Melville, 1851

tiller extension

a wooden, GRP, or carbon-fiber rod attached to the inboard end of a TILLER with the purpose of making it easier to steer while sitting back in the COCKPIT; tiller extensions are usually hinged in such a way that they can be held at any angle to the tiller.

timmey-noggy, a

see GADGET.

tipstaff

a TILLER-like steering arm but set in a vertical instead of horizontal plane; now rarely seen.

toggle

a short piece of wood, usually with pointed ends and grooved around the middle so that it can be fitted through an EYE at the end of a LINE. The toggle is then passed through another eye so that two ropes are secured together but can be quickly released. But anyone who has worn a duffel coat knows that.

toothbrush, half of

The most entertaining part of a recent Whitbread Round-the-World Race (1997–1998) was the decision by one skipper to save weight by demanding that (a) crew members share toothbrushes or (b) that any crew member *really* concerned about oral hygiene should cut most of the handle off his or her toothbrush. Is it not significant that, when a boat performed badly on a leg, the skipper did not demand the jettisoning of more plastic, but demanded the jettisoning of the navigator? Can anyone prove that this sacrifice on behalf of the crew made the slightest difference to the result? And before anyone mentions race psychology, wouldn't the occasional change of clothes have done much more for crew morale?

topmark

a shape, usually of metal construction, fitted to the top of a MARK (usually a BUOY) and used to distinguish the type of the mark.

topmast

the top part of a mast, sometimes a separate SPAR hoisted vertically with a TOPSAIL or the top part of a MAINSAIL.

topping lift

a rope or (more usually) a wire for lifting the BOOM when the sail has been lowered. The topping lift runs from the aft end of the boom to the HEAD of the mast and is adjusted by hauling on the FALL.

topsail (pron. "tops'l")

a topsail is a triangular sail set above a GAFF MAINSAIL, effectively filling the space between the gaff and the main mast; getting one of these up has been likened to launching a huge, very heavy kite; also the second sail above the DECK in a SQUARE-RIGGED ship.

topsides

the part of the hull above the WATERLINE when the boat is not heeling; also everything that is not BELOW DECKS.

torch (US: flashlight)

originally a lantern incorporating a reflector that enables it to throw a beam of light. Now a UK English word to refer to what in US English is a flashlight.

touch and go

to touch the bottom, but get off immediately; a potentially dangerous situation—"It was touch and go there for a moment."

trace

a stainless steel line of high breaking-strain and usually between 1m or 10m in length; used when fishing for bigger fish that might be able and inclined to bite through a nylon monofil line. See TROLLING, FISHING AT ANCHOR.

track

usually defined as "the course a boat travels over the ground." As this could be taken to mean "north on US 101," a better explanation might be "the course a boat travels over the seabed." In other words, the COURSE MADE GOOD (CMG) after allowing for the effects of wind and tidal stream.

track chart

a very large-scale CHART used by ocean passage makers to show the course-made-good over the whole voyage.

Traffic Separation Schemes (TSS)

these are a relatively recent innovation that were introduced when it was discovered that right-of-way rules and common sense together were not enough to prevent collisions in busy shipping zones. Areas such as the English Channel and the approaches to the Panama Canal are now designated TSS with the schemes delineated on charts. ColRegs (US: NavRules) International Rule 10 describes how the seafarer should behave in or near a TSS:

Traffic Separation Schemes

a) This Rule applies to traffic separation schemes adopted by the Organization and does not relieve any vessel of her obligation under any other rule.

b) A vessel using a traffic separation scheme shall:
   (i) proceed in the appropriate traffic lane in the general direction of traffic flow for that lane;
   (ii) so far as practicable keep clear of a traffic separation line or separation zone;
   (iii) normally join or leave a traffic lane at the termination of the lane, but when joining or leaving from either side shall do so at as small an angle to the general direction of traffic low as practicable.

c) A vessel shall, so far as practicable, avoid crossing traffic lanes but if obliged to do so shall cross on a heading as nearly as practicable at right angles to the general direction of traffic flow.

d)
   (i) A vessel shall not use an inshore traffic zone when she can safely use the appropriate traffic lane within the adjacent traffic separation scheme. However, vessels of less than 20 meters in length, sailing vessels and vessels engaged in fishing may use the inshore traffic zone.
   (ii) Notwithstanding subparagraph (d)(i), a vessel may use an inshore traffic zone when en route to or from a port, offshore installation or structure, pilot station or any other place situated within the inshore traffic zone, or to avoid immediate danger.

e) A vessel other than a crossing vessel or a vessel joining or leaving a lane shall not normally enter a separation zone or cross a separation line except:
   (i) in cases of emergency to avoid immediate danger;
   (ii) to engage in fishing within a separation zone.

f) A vessel navigating in areas near the terminations of traffic separation schemes shall do so with particular caution.

g) A vessel shall so far as practicable avoid anchoring in a traffic separation scheme or in areas near its terminations.

h) A vessel not using a traffic separation scheme shall avoid it by as wide a margin as is practicable.

i) A vessel engaged in fishing shall not impede the passage of any vessel following a traffic lane.

j) A vessel of less than 20 meters in length or a sailing vessel shall not impede the safe passage of a power-driven vessel following a traffic lane.
k) A vessel restricted in her ability to manoeuvre when engaged in an operation for the maintenance of safety of navigation in a traffic separation scheme is exempted from complying with this Rule to the extent necessary to carry out the operation.
l) A vessel restricted in her ability to manoeuvre when engaged in an operation for the laying, servicing or picking up of a sub-marine cable, within a traffic separation scheme, is exempted from complying with this Rule to the extent necessary to carry out the operation.

It is a good idea to stay out of Traffic Separation Schemes if you can, but note paragraph (c). If you do have to cross, you are obliged to cross on a HEADING that is 90° to the lanes. I suspect that one's natural instinct is to make a COURSE that is at 90° to them, but the rule says "heading" and so be it.

Traffic Separation Zone (US)

see TRAFFIC SEPARATION SCHEME.

transducer

a THROUGH-HULL fitting used by the echo-sounder (DEPTH GAUGE) to send sound waves toward the sea bottom and receive the resultant echo.

transit

1. the act of being conveyed, especially over a distance.
2. the apparent passage of a celestial object across the meridian of a place.

transom

the flat surface forming the STERN of a boat; a horizontal beam reinforcing the stern of a boat.

traveler

1. a fitting that can be moved to its most effective position, either along a rope or a metal track; most modern yachts have tracks for the mainsheet block and for the jib blocks (one each side of the boat).
2. politically correct name for a GYPSY or Romany person; see OCEAN GYPSIES.

trawler

1. a commercial fishing boat that operates by dragging a scoop-shaped net along the sea bottom.
2. in US English a "trawler" is a motor cruising boat more than 8m (25ft) LOA with a displacement hull and with a lower profile that a sport fishing boat.

Treatise on the Astrolabe

Treatise on the Astrolabe is probably the first ever technical manual written in the English language. It is not remarkable that it explains the use of a NAVIGATION AID (see ASTROLABE); however, it is remarkable that it was written by the English poet Geoffrey Chaucer in 1391, or thereabouts. The document, which includes illustrations, was penned for *Lyte Lowys* or "Little Lewis," a ten-year-old boy who may have been his son or godson. Chaucer was a well-connected diplomat and spy; as neither job involved the use of an astrolabe, the Treatise is thought to be based on other descriptions, especially *Compositio et Operatio Astrolabii* by the 8th-century Arab astronomer Messahala.

In the 14th century documents like this would normally be written in Latin but, as Chaucer pointed out to Lewis, "for Latyn ne canst thou yit but small, my litel sone." Do not be put off by the appearance of the text; the trick is to read it aloud and the meaning will come through clear enough in the sound of it. (The treatise was written long before spelling became an issue and Chaucer thinks nothing about rendering "astrolabe" three or four different ways. It looks as though he texted it to Little Lewis.)

> *Here begynneth the descripcioun of thin Astralabie.*
>
> 1) Thyn Astrolabie hath a ring to putten on the thombe of thi right hond in taking the height of thinges. And tak kep, for from henes forthward I wol clepen the heighte of any thing that is taken by the rewle "the altitude," withoute moo wordes.

2) This ryng renneth in a maner toret fast to the moder of thyn Astrelabie in so rowm a space that it distourbith not the instrument to hangen after his right centre.
3) The moder of thin Astrelabye is thikkest plate, perced with a large hool, that resceiveth in hir wombe the thynne plates compowned for diverse clymates, and thy reet shapen in manere of a nett or of a webbe of a loppe.
4) This moder is dividid on the bakhalf with a lyne that cometh descending fro the ring doun to the netherist bordure. The whiche lyne, fro the forseide ring unto the centre of the large hool amidde, is clepid the south lyne, or ellis the lyne meridional. And the remenaunt of this lyne doun to the bordure is clepid the north lyne, or ellis the lyne of midnyght.
5) Overthwart this forseide longe lyne ther crossith him another lyne of the same lengthe from eest to west. Of the whiche lyne, from a litel cros (+) in the bordure unto the centre of the large hool, is clepid the est lyne, or ellis the lyne orientale. And the remenaunt of this lyne, fro the forseide centre unto the bordure, is clepid the west lyne, or ellis the lyne occidentale. Now hast thou here the foure quarters of thin Astrolabie divided after the foure principales plages or quarters of the firmament.
6) Tlle est syde of thyn Astrolabie is clepid the right syde, and the west syde is clepid the left syde. Forget not thys, litel Lowys. Put the ryng of thyn Astrolabie upon the thombe of thi right hond, and than wol his right side be toward thi lift side, and his left side wol be toward thy right side. Tak this rewle generall, as wel on the bak as on the wombe syde. Upon the ende of this est lyne, as I first seide, is marked a litel cros (+), where as evere moo generaly is considerid the entring of the first degre in which the sonne arisith.
7) Fro this litel cros (+) up to the ende of the lyne meridionall, under the ryng, shalt thou fynden the bordure divided with 90 degrees; and by that same proporcioun is every quarter of thin Astrolabie divided. Over the whiche degrees there ben noumbres of augrym that dividen thilke same degres fro 5 to 5, as shewith by longe strikes bitwene. Of whiche longe strikes the space bitwene contenith a myle wey, and every degre of the bordure conteneth 4 minutes, this is seien, mynutes of an houre.
8) Under the compas of thilke degrees ben writen the names of the Twelve Signes: as Aries, Taurus, Gemini, Cancer, Leo, Virgo, Libra, Scorpio, Sagittarius, Capricornus, Aquarius, Pisces. And the nombre of the degrees of thoo signes be writen in augrym above, and with longe divisiouns fro 5 to 5, dyvidid

fro the tyme that the signe entrith unto the last ende. But understond wel that these degres of signes ben everich of hem considred of 60 mynutes, and every mynute of 60 secundes, and so furth into smale fraccions infinite, as saith Alkabucius. And therfore knowe wel that a degre of the bordure contenith 4 minutes, and a degre of a signe conteneth 60 minutes, and have this in mynde.

9) Next this folewith the cercle of the daies, that ben figured in manere of degres, that contenen in nombre 365, dividid also with longe strikes fro 5 to 5, and the nombre in augrym writen under that cercle.
10) Next the cercle of the daies folewith the cercle of the names of the monthes, that is to say, Januarius, Februarius, Marcius, Aprilis, Maius, Junius, Julius, Augustus, September, October, November, December. The names of these monthes were clepid thus, somme for her propirtees and somme by statutes of lordes Arabiens, somme by othre lordes of Rome. Eke of these monthes, as liked to Julius Cesar and to Cesar Augustus, somme were compouned of diverse nombres of daies, as Julie and August. Than hath Januarie 31 daies, Februarie 28, March 31, Aprill 30, May 31, Junius 30, Julius 31, Augustus 31, September 30, October 31, November 30, December 31. Natheles, all though that Julius Cesar toke 2 daies out of Feverer and putte hem in his month of Juyll, and Augustus Cesar clepid the month of August after his name and ordeined it of 31 daies, yit truste wel that the sonne dwellith therfore nevere the more ne lasse in oon signe than in another.

Here endith the descripcioun of the Astrelabie.

Well, not quite. There is much more and competitive translation is a fun after-drinks game while riding at anchor waiting for the foul weather to blow through. If you get good at that, take up the challenge of trying to draw an astrolabe from this description; then compare the result with the photograph on the British Library website (also on Wikipedia). When you study the photograph, look out for the little dog! You can download the rest of the text from Google Books or various other sites. The document is, however, unfinished and this brings me to the sad ending to the tale; Little Lewis died before Chaucer could finish this elegant piece of instruction.

triatic stay

a line, usually made of wire, that connects the heads of the two masts of a FORE-AND-AFT VESSEL.

trim

1. the trim is the FORE-AND-AFT, port-and-starboard aspect of a vessel when she is afloat.
2. to trim: to adjust the sails so that they draw most effectively.

trimaran

A yacht with three HULLS, the larger, central one providing most of the ACCOMMODATION while the smaller, outboard hulls provide lateral buoyancy. Alternatively, a yacht with a hull so narrow that that it needs floats at each side to reduce the amount of HEEL she experiences. A friend of mine has a trimaran whose outer hulls fold up so she becomes towable.

tripping line (US: trip line)

a line attached to the crown of an ANCHOR in such a way that it can be used to BREAK OUT the anchor from the seabed; such a line can run back to the BOW of the boat or to a BUOY.

trolling

fishing by pulling a lured hook behind a boat.

Many yachtsmen carry a sea fishing rod and tackle, but it is difficult to use these when under way. The solution is to "troll" with a purpose-built rig. Trolling is based on the principle of dragging a lure behind the boat and has the advantage that you can be fishing all the time you are sailing. I have seen some kits for fishing from sailboats on sale in chandlers' but have been impressed with none of them. The solution is to find the local sea anglers' store and explain to the person running the shop what you want to do. Tell them you want to troll for tuna off the stern of a yacht. What he should offer you is a variety of hooks and lures (we will come to those later) and a selection of steel traces at least 50cm long (big fish have teeth that will go straight through any fishing line); heavy-duty swivels (to stop the line tangling) and line. There are two types of line: nylon monofilament ("mono" or "monofil") and braided terylene. Monofil stretches; braided line doesn't. Braided seems to be more resistant to sudden breakage if it becomes chafed. However, mono is more likely to take the strain of a "bite" and this is crucial if you are not using a rod; hence that is my preference. You need line with a breaking-strain of 80lbs ("80lb test"), 36kg, and if you can afford it, buy an 800m drum of the stuff. This will enable you to make backup lines. The shopkeeper might offer you a wooden frame onto which you can wrap the line as you haul it in; make sure it will hold at least 300m.

Here's how you rig the line. Securely fix the lure to the hook (unless the lure has a hook built into it already). Then attach the hook to the trace (via a strong swivel) then the trace to the line (via another swivel—no such thing as too many swivels). Now decide where you are going to secure the inboard end of the line. A stern cleat is the best solution, but *make sure the line will run out without chafing against any other part of the boat*. Keep in mind that the rig might be out there for some days. Now feed the lure out—about 200m should do it. Check the line periodically by pulling it. If it feels heavy—and it will if there's a fish on it!—start hauling in. It is just hard work from this point; this is not sport, it is food gathering (wear gloves; if necessary use the winch).

You can add a little sophistication in the shape of a "bite-detector." Improvise a short pole (called an "outrigger") that will reach out from the transom by a meter or so (you need to be able to reach the end). It could be lashed to the pushpit. At the outboard end, fit a strong clothes-peg (clothespin in the US). Once you have deployed your line and secured it to the cleat, pull in a BIGHT and feed it though the peg. With luck it will hold there until the lure is taken by a fish, when the line will snap free. If the peg doesn't hold, reinforce its jaws with rubber bands or shock cord.

Here is an alternative arrangement. Some sea anglers—especially in Britain—use old-fashioned (i.e., low-tech) wooden reels. These can be rigged horizontally on a vertical pole at the stern. A length of elastic shock cord can then be rigged around the reel and back to a fixed point to stop the reel running out and to take the initial shock of the bite.

If you are handy with wood, you might even be able to make one of these reels yourself. And if you are going really low-budget, you can improvise a bite alarm. Wash out a small baked bean can. Place a few odd nuts and bolts inside and tape the lid down. Place the can where the movement of the boat will not dislodge it and connect it to the shock cord with a length of light line. Proceed with normal shipboard life until you hear the can crash onto the deck. Retrieve your leviathan.

And talking about leviathans. . . . If you leave a fish on the lure for too long, it becomes bait for even bigger fish. Like sharks, for example. What you are most likely to retrieve in this situation is a head with a surprised expression, but sharks are known to take the whole fish, hook and all. So don't be too surprised if you seem to be pulling in a Volkswagen Beetle.

see KILLING FISH; FISHING AT ANCHOR, BUYING FRESH FISH.

tropical storms

powerful rotating storms that occur in tropical and subtropical climates. They circulate in a counterclockwise direction in the Northern Hemisphere and clockwise in the Southern Hemisphere. The "eye" usually moves at speeds of about 25 KNOTS, but the package arrives with wind speeds in excess of 64 knots, sometimes well over 100 knots. No haven is safe from hurricanes and even having the boat on the HARD will not guarantee its safety. The names given to tropical storms vary throughout the world, but they are all offer the same experience if you are in or near one:

Hurricanes	Western North Atlantic Eastern North Pacific Western South Pacific
Cyclones	Arabian Sea Bay of Bengal Southern Indian Ocean Northwestern Australia
Typhoons	Western North Pacific

The next time you are bored out of your mind watching ice skating on television, take a close look at the way the performers spin. Once they have gained enough momentum to turn on the spot, they draw in their arms and legs. This makes them spin faster. It works the same way with a tropical storm. As the depression deepens, it gets smaller and smaller and, consequently, rotates faster and faster.

Tropical storms are real killers but can be avoided through good passage planning. Tip: If you are unfortunate enough to have a tropical storm on your tail, head for the equator. Hurricanes rarely encroach within 8° north or south of the equator. Otherwise restrict your sailing to the South Atlantic. They do not occur there at all.

See TYPHOON.

trough

the "valley" between two waves.

trysail (pron. "trys'l")

a small but very rigid triangular sail used only in heavy weather conditions; usually set to the MAINMAST instead of the MAINSAIL.

tsunami

a huge powerful wave mostly occurring in the west Pacific. Tsunamis are caused by undersea earthquakes and are not "tidal waves." I understand that *tsunami* is Japanese for "village wave," probably because it wipes out your village.

tuck

"tucking" is the process of weaving the open strands at the end of a rope to make a splice.

tug

a powerful vessel used to assist the maneuvering of ships in confined waters; tugs are designed to tow or push. Most tugs are HULLS wrapped around very big engines, in other words they have a very high power-to-weight ratio. Huge sea-going tugs are often seen towing oil rigs; at other times they wait in strategic places such as Gibraltar, monitoring VHF Channel 16 for high-value salvage opportunities.

tumble-home

expression used to describe the inward curve of the upper sides of a ship's hull; the opposite of FLARE.

tuna

the staple protein supply for the ocean cruiser in tropical waters.

In 1992, a 340kg (784lb) bluefin tuna sold for $83,500 in the Tokyo fish market. So, if you get a specimen like that and you are not too far from Japan, try and find a way of keeping it fresh. That or get on the VHF and try to raise a passing fishing boat that will do business. See TROLLING.

tuna tower

a tall (usually aluminum) tower on a sport fishing boat; used for spotting fish in the distance and often equipped with a duplicate set of controls.

turnbuckle

screw-threaded device used to adjust the tension in stays and other parts of the standing RIGGING; see also BOTTLESCREWS.

turning blocks

BLOCKS used to change the direction of a line in the running RIGGING.

turn up, to

1. to be on the pier-side when you promised.
2. to make a line fast to something.

turtles

This guide to the sea turtle is based upon the experiences of the Robertson family after their 43ft schooner *Lucette* was struck by a killer whale and sank in 60 seconds. This happened in June 1972 and I suspect Dougal Robertson's account of how he, his wife Lyn, elder son Douglas, twin sons Neil and Sandy, and an American student called Robin survived for thirty-eight days is still the best textbook available on survival at sea. Turtles played a big part in the satisfactory outcome but it is worth noting first of all that the Robertsons escaped from *Lucette* with two boats, a dodgy out-of-date liferaft (which sank first), and a 3m GRP tender. This is significant because it would be significantly more difficult to kill and butcher a turtle in an inflatable liferaft.

However, first catch your turtle. Turtles bite and their powerful claws are sharp—so care is needed. Smaller ones can be manhandled over the side but in the case of larger ones it might be necessary to lasso a flipper or two. According to Dougal, this is what you do next.

> Once aboard and laid on its back a turtle is helpless but not harmless; determination and a sharp knife to the turtle's throat soon turns the situation in the castaway's favour but it is essential to ensure that the main blood supply is tapped to achieve speedy results. The collection of the blood is quite easy if the spurting blood vessel can be directed into a cup but the blood must be drunk immediately for it coagulates in about thirty seconds.

Unappealing as it may seem, the turtle's blood probably saved the Robertsons from dying of thirst (they had very little water). After prying open the undershell, they would eat the flesh—but avoided the liver, which is thought to be risky to eat. *Survive the Savage Sea* by Dougal Robertson is highly recommended (even the Hollywood movie of the same title wasn't too bad).

typhoon

a tropical storm. A slight distortion of the Cantonese *dai feng* (pronounced something like "die fung"), and meaning "big wind." See TROPICAL STORMS.

unbend

to remove a sail from the MAST, BOOM, and running RIGGING.

under bare poles

see BARE POLES.

under-run, to

a method of WEIGHING a reluctant KEDGE anchor. Maneuver the TENDER under the WARP (RODE), placing it in the sculling notch, if you have one. Haul away so that, as you pull in the line, the dinghy gets closer and closer to the ANCHOR. Once you are over the kedge, keep pulling and the anchor will come up. Either that, or the dinghy will sink.

under way

a vessel is said to be "under way" when she is moving through the water and will answer to the helm.

Universal Time (UT)

see GREENWICH MEAN TIME.

unreeve

to pull a line out of a BLOCK or EYE.

unstayed vs. stayed rigs

an unstayed rig is one in which the mast or masts of a vessel are free-standing; there is no visible means of support.

Anyone who has seen or sailed a lively Laser dinghy will be familiar with an unstayed rig. Its appeal is in its fundamental simplicity; short of a failure of the mast itself, there is nothing that can go wrong with it. So why do most dinghies and yachts have masts that can only be kept up by some combination of FORESTAY, forward lower SHROUDs, cap shrouds, aft lower shrouds, single, double or Y BACKSTAYs along with bottlescrew DECK fittings, lever or power tensioners, CROSS-TREES, and various other mast fittings? Particularly if they are intended for cruising where reliability (and, arguably cost) are a higher priority than performance? It seems to me that the design of sailing craft could benefit from a fundamental rethink of the way in which we convert wind to motive power; just because most boats are Bermudan-rigged does not mean that is the perfect way of doing it.

See JUNK RIG. See also BERMUDAN RIGS and LUG RIGS.

up and down

when an anchor is being weighed, its WARP (US: RODE) is said to be "up and down" when it is all hanging vertically and, therefore, the ANCHOR is about to break out.

up helm

pulling the tiller (or turning the wheel) so that the boat turns off the wind (BEARS AWAY).

vane

see WIND VANE.

vane self-steering gear

a STERN-mounted arrangement that allows a sailing boat to follow a course in relation to the direction of the wind.

The first reliable version of this system was developed by the famed single-handed racer Blondie Hasler. The wind cannot be relied on to be strong enough to turn a RUDDER. The trick, therefore, was to use the wind to *collapse* (not turn) a vane projecting above the self-steering rig. That provided enough leverage to turn a vane (or tab) in the water, thereby letting the flow of water provide the force to turn the rudder. Once the boat was back on course, the wind vane would pop back up again. Very clever. Blondie Hasler was also a very courageous man, being decorated for his bravery as a "Cockleshell Hero" during World War II. He should also have been decorated for his promotion of the JUNK RIG.

vang, boom vang (US)

a rope that controls the head of a gaff boom; but see also KICKING STRAP.

Vanikoro shipwreck

This is the last of three entries that retell the story of a 1930s passage from Denmark to the Pacific Ocean. Part 1 is under MONSOON and Part 2 is under GALAPAGOS ISLANDS.

The story of the ketch *Monsoon* and its crew of relentlessly cheerful Danes ends in the western Pacific. They are anchored in the lagoon of Vanikoro in the Santa Cruz Group. The place does not appeal and they decide to move on. Hakon Mielche gives us a graphic description of what happened next.

> When we were still five hundred yards from the entrance there came the most unexpected fierce squall with fearful blasts of wind and heavy rain. The waters of the lagoon were whipped into a foam in a second, and several of us rushed on deck to see what was happening. Even below deck we had heard the clap with which the wind and the rain had struck against the ship and forced it to starboard. It was as though a giant had hit the *Monsoon* with his fist.
>
> The squall passed as quickly as it had come and everybody breathed a sigh of relief. It was rotten luck that it came at all, but a good thing that it did not happen a little later when we were in the middle of the passage. [It is often better to be mid-passage, in open water, well clear of a lee shore when a squall hits. JL]
>
> Sonny and I went forward to keep an eye on the channel. Mr. Möller stood aft beside the mate and Kalle, who had the wheel; Jack stayed amidships, while the skipper was still aloft, from where he gave his orders to the man at the wheel. The wind was slightly stronger after the squall had passed. The wheel was put over a shade to port, so that we would be as far as possible on the weather side when we got into the narrow channel through the reef. The next squall came when we were five yards away from the reef, more unexpected and more violent than the first. Then events came tumbling over each other like strips of flickering film put together without rhyme or reason and wound off in mad haste.

This story makes me appreciate weather forecasts with synoptic charts downloaded via shortwave radio or satellite and the critical decision: go or stay? If in doubt, stay. But it seems that the crew of the *Monsoon* was tragically committed.

> The whole world was blotted out in a grey mist in which were only the *Monsoon*, the sea, the storm and the coral reef shining hungrily through the light green of the water to starboard.
>
> Nearer and nearer came the reef, like in a nightmare when the most terrible dangers came slowly closer without one being able

> to move hand or foot. Through the foam we could clearly distinguish the shape and colour of the individual rocks and where the reef plunged steeply into the depths turning a sharp, jagged edge towards us. The motor hammered at full speed. We accelerated as much as we could, but the thirty-five horse power was not able to do much against the storm. We were helplessly, slowly but surely being forced sideways to destruction. For every inch that we went forwards, we were driven ten to the side. We could do nothing. We were the helpless, bewitched spectators of a play which we knew must end in tragedy unless a miracle happened.
>
> The skipper had long since ordered the wheel to be put hard over to port. The man at the wheel had obeyed almost before the order had been spoken—but the *Monsoon* refused to answer. No power on earth could have forced the stem to windward against the pressure of the mad gusts which howled in the shrouds and rigging. The rudder was a useless piece of firewood. Nearer and near came the reef—twenty yards—fifteen—ten—seven! We were in the channel. There was no turning back. Even if the ship had obeyed the helm, there was not room to turn.
>
> "Hoist the mainsail and ease the helm!" roared the skipper from his lookout, and in the same instant every man on board hurled himself at the braces, the helmsman eased the wheel to starboard, and the mainsail slid up with a speed unknown in the old ship's glorious history. Fourteen hands held and tugged with all their might, sailors, owner, scientist and cook, they all held on for their lives and for the sake of the old boat; but there was nothing to be done.
>
> Before the sail was half up, before there had been the slightest possibility of easing off and trying this last remedy, the *Monsoon* went aground for the first time with a bump which cut us to the quick, and before the sail was furled the breakers had thrown us onto the reef.

Except when swimming I've never been on a reef, but this description gives me a hollow feeling in the pit of my stomach. They were in one and a half fathoms and it was high water on a spring tide. The *Monsoon* was finished, but could the crew save themselves?

> There was no panic. In fact, apart from a quivering feeling of tension and seriousness, we felt nothing. Orders followed one another in rapid succession and we obeyed without an instant's delay. The breakers thundered against the port side; the keel kept bumping against the sharp hard coral; and yet the motor-boat was into the water like lightning and the mate went with Jack and Kalle to try to drop both anchors in the deep water of the passage, one fore and

one aft, which was the only way left of keeping the ship on the edge of the reef and near deep water, the last vague chance of getting her off again when the storm had subsided and the seas slackened. They succeeded in putting out both anchors, but on the way back the motor-boat was slung against the ship's overhanging stern and it was only by pure chance that Jack was not crushed against the planks.

Everybody was in the cabins in all haste packing their valuables and those of the ship. It was only a question of time before she sprang a leak and the water would rise inside. To be sure, the *Monsoon* was a miracle of oak planks and strength, but a coral reef is hard and more merciless than anything else in the world. The squall was over, but the wind still blew as hard as ever. The seas crashed thundering against the ship's side and poured continually over the deck, the lagoon was in a state of commotion, the foam flying from the tops of the waves like the manes of white horses, and the whole vessel lay on its port side so that we had to hold on to the rigging and booms if we wanted to cross the deck without falling.

The motor-boat bumped against the planks and Jack sat in her working at the engine, wet to the skin. There had been no time to put on oilskins; but in such a situation you do not notice that you are wet. Sacks of clothes, chests of nautical instruments and bundles of papers were thrown into the boat; then Stubbe and Jan jumped in, the painter was cast off and the motor began its struggle against the foaming seas.

from *Let's See If the World Is Round* by Hakon Mielche, 1944
(originally published in Danish in 1938, now out of print)

Thankfully, all our intrepid Danes (and Bobby the Dog) managed to get ashore, but *Monsoon* was lost and their great adventure was over. Such was their luck that they got back to Denmark just in time for the German invasion of April 1940.

variation

the difference between true north and magnetic north; not to be confused with DEVIATION. When God designed the world he included magnetic variation to make life difficult for seafarers; magnetic north would never be the same as true north and, just to make it even trickier, it would become a moving target. Variation, therefore, is shown on CHARTS (in the COMPASS ROSES) and changes year by year. This difference is quite important to would-be polar explorers; magnetic north is sometimes in Canada within easy sledding distance of Yong Street, Toronto.

Vaseline®

see PETROLEUM JELLY.

v-berth

the double-berth arrangement in the forecabin of a boat; here it is that you get to play footsie with another member of the crew or with a heap of wet sailbags.

vectors

forces can be represented as numbers and manipulated mathematically or they can be represented as lines on a piece of paper and manipulated graphically; in this case they are called vectors and are at the heart of basic chart navigation. So, the desired track of a boat can be represented by one line showing its course from departure point to destination point. The effect of the tide for the period of that journey can be represented by one or more vectors representing the direction of the current and the distance it will move. A third line from the end of the tide vectors to the destination will provide the course to follow to counter the effects of the tide. Very satisfying when done correctly.

veer

1. a clockwise change in wind direction; for example, when the wind direction changes from 180° (South) to 270° (West), it is said to have "veered." See also BACK.
2. to pay out ANCHOR cable.

ventilation

keeping the air moving when anchored-up in tropical climes can be an important contribution to comfort (and one's ability to sleep). One of the most effective and economical solutions is a tent-like air-scoop (at a cost of about £40 or $60, the Windscoop™ is a good example). This is suspended over the forward hatch using a halyard. The open "door" points in the general direction of the wind, so encouraging the air to move down into the forecabin, along the saloon, and out through the main COMPANIONWAY.

Another possibility if there is no wind is to buy one of those little "personal fans" designed for use in cars. They plug into a standard LIGHTER SOCKET.

Vertue

a delightful pocket cruiser based on the FOLKBOAT.

> Let them recognise virtue and rot for having lost it.
>
> Aulus Flaccus Persius (32–62 AD), Roman satirist

Very pistol

a large-caliber hand gun used to fire signal flares; originally 12-gauge (UK: 12 bore) but now standardized internationally on 25mm. Sometimes called a flare pistol. See FLARES.

VHF

Very High Frequency radio; see RADIO.

victuals (pron. "vittles")

food.

wake

the trail of disturbed water left by any vessel as it moves through or over the water; a DISPLACEMENT boat leaves two wakes, one from the BOW, one from the STERN.

Walker Log

a LOG that is trailed from the TAFFRAIL or PUSHPIT; a means of measuring a boat's distance traveled.

warp

1. any line used when WARPING.
2. a line (rope or cable) used to secure a vessel when in harbor.
3. a rope attached to a KEDGE anchor is properly called the "kedge warp"; now commonly used to refer to any line connected to any ANCHOR.

warping

to move a vessel in harbor by pulling on lines secured to ANCHORs and/or bollards. A means of inducing great terror in the skippers of neighboring yachts (even though "warp-speed" in this case is considerably less than that achieved on the Starship Enterprise).

washboards

wooden boards that slide vertically (or near-vertically) into grooves at each side of the COMPANIONWAY to prevent water getting below. They also provide a secure enclosure for the boat when she is unattended. These things are practical, but a nuisance at the same time. When they are in place, they can make it hard work going between the accommodation and the cockpit. Take them out and they get under your feet.

washing machine

For a quick sniff of what biological warfare must be like, slide back the hatch of an ocean racing yacht shortly after it has crossed the line. Twelve men and women in such a confined space wearing the same clothes for weeks on end. (See TOOTHBRUSH.) Even a card-carrying character like me enjoys the occasional change of clothes, so why not try Lagan's un-Patented Lo-Tech Washing Machine? No, not the bucket, the plastic refuse sack. The ones to buy are black and not safe for children (for obvious reasons). Chuck in the clothes. Add about the right amount of washing powder and a few liters of fresh water. Tie a tight knot in the top making it as watertight as you can. Add tape if necessary. Now place the sack somewhere where it will benefit from the motion of the boat. I use the cockpit. In tropical latitudes, the black bag will absorb the heat of the sun. Set for the appropriate cycle. Make a coffee, switch on some "easy listening" and settle back with a good book. Let it rock 'n' roll all day. Then rinse with clean fresh water and hang out to dry. Incredible but effective! Do not mix up this bag with the real trash. Do not mix colors and whites, boys.

watch

system of shifts used to ensure twenty-four-hour cover on a vessel; traditionally organized in the following way:

Start	End	Watch	Day 1	Day 2
0000	0400	Middle	Port	Stbd
0400	0800	Morning	Stbd	Port
0800	1200	Forenoon	Port	Stbd
1200	1600	Afternoon	Stbd	Port
1600	1800	First Dog	Port	Stbd
1800	2000	Last Dog	Stbd	Port
2000	2400	First	Port	Stbd

The example shown is based on two watches, traditionally called "port" and "starboard." With three watches ("red," "white," and "blue") it becomes easier to arrange eight-hour sleep periods and that was an innovation of that great man-manager Captain James Cook. A "dog watch" occurs when all members of the crew are awake and on DECK at the same time. Not only does this make it possible to rotate crew through the watches, it also makes it possible for cook to serve one main meal a day.

Other factors that need to be taken into account include the comparative experience and skills of the watch members. None of the above should be confused with the "DECK WATCH" which is a timepiece used in celestial navigation.

water, water

fresh water is one commodity that is immensely difficult to do without when passage making. For centuries provisioning was undertaken by loading a vessel with wooden casks that, because of their weight, would be stored as low down in the hull as possible. At each port of call, empty casks would be refilled; full casks would also be carried up on DECK, washed out, and refilled. This was a physically strenuous operation that resulted in injury for many of the crew. Casks were eventually replaced with metal tanks and pumping systems.

The simple turn of a tap in the galley of a cruising boat should provide a drinkable glass of water. On a modern yacht that will come from a pressurized plastic or rubber tank (or series of tanks). The seafarer will probably have bought the water itself in the last marina or harbor visited. The supplier might have been some guy with a little tanker (like a fuel BOWSER) who will be paid by the liter or gallon. The flashier marinas will have a metered tap close by each berth, supplying "drinking water." There may also be an outlet offering "fresh water." Do not mix these up; the latter will be salt-free and suitable for washing-off decks and rigging to free them of accumulated salt after a voyage; it is not necessarily safe to drink.

The Big Question here is: Should any of these sources be trusted? The old guy with the tanker may be supplying you with salt-free water, but what is it *not* free of? Does the water coming out of the metered taps on the dock come from a fresh-water lake or stream? Does it come from a well? When did the marina's manager last have the water laboratory-tested and can you see the results, please? In some havens the water will come from a REVERSE-OSMOSIS plant (more on which later). Even if the water is well purified, what condition is it in by the time it has gone through a network of poorly maintained pipes to reach your berth? (If using the same hose to spray fresh water onto the boat and to pump drinking water into your tanks, remember to flush the fresh stuff out with some of the drinking water before sticking the hose into the through-deck filler cap.)

It is always a good idea to buy some bottled drinking water to supplement your tank supply. There are two reasons for this: if, five hundred miles out into the Pacific, you discover that the tank water is contaminated, you have a serious crisis; secondly, if you have to abandon ship, you can take some of the bottled water with you in your LIFERAFT. Here we are not talking about tubby little green bottles of Perrier designer water, but huge five-liter-plus plastic containers (*still*, not bubbly!). Once empty, these can be refilled if a source on shore is found. Warning: I have seen rows of these plastic containers secured to the stanchions. If you are going to do this, pieces of sponge should be used to limit the chafing—the plastic is usually very thin.

Being surrounded by huge quantities of water and being short of the stuff you can safely drink is very frustrating. The original solution to this—500 years ago—was to employ stills. These worked on the principle that boiling salt water would cause the H_2O to vaporize off, leaving the salt behind. A copper coil would then be used to convert the vapor back to liquid water again. The Royal Navy refused to have such stills onboard ships because they can be readily used to distill alcohol. Some boats had them, though.

> After that, one of them part filled the boiler with fresh water, and soon we were occupied most pleasantly upon a supper of cold, boiled salt-meat, hard biscuits, and rum mixed with hot water.
>
> During supper, the bo'sun made clear to the men regarding the watches, arranging how they should follow, so that I found I was set down to take my turn from midnight until one of the clock. Then, he explained to them about the burst plank in the bottom of the boat, and how that it would have to be put right before we could hope to leave the island, and that after that night we should have to go most strictly with the victuals; for there seemed to be nothing upon the island, that we had up till then discovered, fit to satisfy our bellies. More than this, if we could find no fresh water, he should have to distill some to make up for that which we had drunk, and this must be done before leaving the island.
>
> *The Boats of the "Glen Carrig"*
> by William Hope Hodgson, 1757

For today's solution: see REVERSE OSMOSIS.

The only answer for a cruising boat running out of drinking water (something that should really never be allowed to happen), is to look for land and pray that is has potable water.

Be that as it would, we were obliged to go on shore somewhere or other for water, for we had not a pint left in the boat; when or where to get to it, was the point. Xury said if I would let him go on shore with one the jars, he would find if there was any water, and bring some to me. I asked him why he should go? Why I should not go and he stay in the boat? The boy answered with so much affection, that made me love him ever after. Says he, "If wild mans come, they eat me, you go way." "Well, Xury," said I, "we will both go; and if the wild mans come, we will kill them, they shall eat neither of us." So I gave Xury a piece of rusk bread to eat, and a dram out of our patron's case of bottles which I mentioned before; and we hauled in the boat as near the shore as we thought was proper, and so waded on shore, carrying nothing but our arms and two jars for water.

I did not care to go out of sight of the boat, fearing the coming of canoes with savages down the river; but the boy seeing a low place about a mile up the country, rambled to it; and by and by I saw him come running towards me. I thought he was pursued by some savage, or frighted with some wild beast, and I ran forward towards him to help him; but when I came nearer to him, I saw something hanging over his shoulders, which was a creature that he had shot, like a hare, but different in color, and longer legs. However, we were very glad of it, and it was very good meat; but the great joy that poor Xury came with was to tell me he had found good water, and seen no wild mans.

But we found afterwards that we need not take such pains for water, for a little higher up the creek where we were we found the water fresh when the tide was out, which flowed but a little way up; so we filled our jars, and feasted on the hare we had killed, and prepared to go on our way, having seen no footsteps of any human creatures in that part of the country.

Robinson Crusoe by Daniel Defoe, 1719

See also CANNIBALISM.

waterline

the line where the HULL and the surface of the water intersect.

wave

a moving or undulating ripple on the surface of the sea; the origins seem complex, but there are Old English words *wǽg* and *wagian* meaning to shake or to wag as in "wagging tongue." Middle English has the noun *wawe*, a wave.

Waves are interesting because they constitute a large proportion of our environment when at sea; after all, apart from the weather and the night sky, there isn't much else in which to take an interest. Waves are caused by the passage of air across the surface of any liquid. The unhindered wind blowing across the flat surface of the open sea "picks up" the surface molecules of water, which show as tiny ripples, and the sea begins to get lumpy. When the wind falls, the waves subside again. The wind blowing over the surface of the sea is called a "disturbing force." Other types of wave have different "disturbing forces": sub-sea volcanic eruptions and earthquakes give rise to TSUNAMIS. Waves themselves can be the disturbing cause of other types of wave. For example, a "seiche" wave is the consequence of powerful waves entering an enclosed area, such as a harbor or nearly enclosed bay, and sloshing about once trapped.

A "surface wind" disturbing force picks up small waves that are known as "capillary waves." This process starts at about two knots. As the wind quickens, it is able to get a better grip on the rougher surface and turns the small waves into bigger waves. From the seafarer's point of view the key element in all this is that the motion of the waves is in the same direction as the wind. Keep in mind that it is the *waves* that are doing the traveling, not the water; the motion does the moving. To convince yourself of this, take a length of line and flick it like a whip. Waves will travel along the length of the rope, but at no time will the rope leave your hand. Similarly, a fishing float will rise and fall as a wave passes under it.

Below the waves, the water is tumbling in counter-rotating somersaults. This effect (known as "orbital motion") only lasts for a few meters and, the deeper the water is, the less effect the waves have on the motion of the sea at the bottom. You can work this out; the circular motion of the waves stops at about one-half of the wavelength (see below). So, if the wavelength on the surface is ten meters (30 ft), the water will be decreasingly turbulent down to five meters. Eventually, when the wind speed gets up to eleven knots, it starts to rip the crests apart, creating "whitecaps" or "white horses." This is the point at which you can determine that you are experiencing Force 4 on the BEAUFORT SCALE. From seventeen knots (Force 5) there are even more whitecaps and spray is being torn off the top of the crests.

It is useful to be familiar with the names used for the different parts of a wave and with how a wave is measured. The tops of waves are known as the "crests" and the deepest parts between the crests are known as the "troughs." The "wavelength" is, as the word suggests, the distance between a crest and the preceding or following crest. The distance between adjacent troughs should, of course, be the same. "Wave height" is the vertical distance between the trough and crest of a wave and is determined by the strength of the wind and by the "fetch"—the distance the wind has been blowing over clear water. A very approximate indication of maximum wave height in feet is given by dividing the wind speed in miles per hour by two; a 50 mph wind results in 25 ft waves. "Amplitude" is a technical term that you will only hear oceanographers use; the same applies to the "still-water line," a notional concept and something that is virtually impossible to estimate in real life. The "wave period" is the time in seconds it takes a complete wave (crest-to-crest or trough-to-trough) to pass a fixed point. As you are unlikely to be perched on a fixed point, this is not an easy measure to estimate. Related to the wave period is the "wave frequency," the number of waves that pass a fixed point in, say, sixty seconds. Neither of these values are much use in practical sailing other than to observe that the smaller the period, the higher the frequency, the more uncomfortable will be your ride.

The shape of waves provides clues to their origin. For example, short, choppy seas offshore are usually caused by a recent storm in the area. Longer, steady waves with high crests are likely to be much older and may have traveled hundreds of miles. Most people see waves when they finally crash out against a shoreline, maybe even onto a beach. As the turbulent underside of waves comes into contact with the upward-sloping bottom of the ocean, the front part (or "face") of the wave is made to slow down. This causes the back part of the wave to catch up and try to climb over the front, increasing the height by as much as fifty percent. The resulting, *moving*, build-up of water curls over at the crest and crashes onto the beach. This form of wave is called a "breaker," or "surf." When the water from the waves rushes back into the sea, depending on the shape of the shoreline, dangerous "rip currents" can be formed. A shallowing bottom can also cause waves to change direction in a process called "wind refraction." The line of the waves will turn so that they are more parallel to the shore.

See SWELL, SEVENTH WAVE.

way

when a boat is "under way," she is moving through the water.

waypoint

the latitude and longitude coordinates of a specific location being used for navigation purposes; in GPS receivers these may be named, numbered, and even downloaded from the Internet.

wear

to go round STERN to wind from the STARBOARD tack to the PORT tack and vice versa instead of going through the wind (head to wind) as in TACKING.

weather, to

to pass to WINDWARD of any obstacle in a sailing boat without having to TACK.

weather eye

It is a good idea to "keep a weather eye open"; in other words, look toward the "weather" side of the boat—the side that the wind (and, invariably) trouble will come from. Except when trouble is in the form of a LEE SHORE, of course.

weather helm

a yacht has weather helm when the HELM (or wheel) has to be held to WINDWARD (of MIDSHIPS) to maintain a course. This usually means that the yacht is over-canvased and requires REEFING.

weather shore

the one that lies to windward and is infinitely preferable to a LEE SHORE; however, if the weather is coming from where you are trying to arrive you might have a different view.

weather side

the side of a vessel that is closest to the wind.

"weather, under the"

being battered by a heavy wind, in nautical terms; today used in everyday English to mean "feeling ill" (i.e., suffering from a hangover).

weigh

to weight an ANCHOR, to raise an anchor.

whack

one crew member's daily ration (obsolete); gave rise to the expression "a fair whack," a reasonable share.

wheelhouse

see PILOT HOUSE.

whip

1. to whip: process of securing the open strands at the end of a rope by tightly binding with twine, a fine craft with fancy variations.
2. a whip: a single-BLOCK PURCHASE.

whipping

the result of binding the ends of a rope to prevent the strands from unraveling.

whisker pole

a long, narrow spar fitted to the mainmast at one end and the CLEW of the FORESAIL at the other when RUNNING downwind.

whisky (whiskey)

whisky (spelled "whiskey" in Ireland and the US) is an alcoholic drink made from fermented corn barley or rye; from the Scots Gaelic *usque beatha*, "water of life." Of particular interest to canal pilots, customs officers, and various port officials who in many parts of the world claim authority to check the quality as well as the quantity of your liquor stores; rarely a good idea to refuse. Of this motley bunch of characters, only the revenue men (customs) are empowered to collect duty on your stock. (The others will collect the unfinished bottles if you are lucky.) See also rum, gin, vodka, brandy, quality wines, cigarettes, cigars, pipe tobacco. . . .

wim-wom

see GADGET.

whole nine yards, the

all squaresails set. I'm not entirely convinced but the nautical version goes something like this. A common sail-plan for a square-rigger would have three main masts plus a MIZZEN mast. Excluding the last, each mast would have three YARDS. So, fully rigged, the ship would have nine sails set on nine yards and would be good to go.

winch

a mechanical device enabling purchase to be applied to lines under strain; consists of a drum containing a gear box operated by a removable handle on top of the winch; the line is turned around the drum and can be pulled until the strain demands the use of the handle; some (more expensive) winches are electrically or hydraulically powered; other types are "self-tailing" to make single-handed operation easier; the winch has largely replaced the use of blocks on modern yachts; mostly used for SHEETS and HALYARDS.

windlass

a large, powerful WINCH fitted with a GYPSY and used primarily for raising and lowering the ANCHOR; originally worked by hand or steam, now also by electricity or hydraulics.

wind rose

a graphic device shown on pilot (passage-planning) charts to indicate the strength, direction and frequency of the wind for a particular place and at a particular time of year; essential for long-range passage planning. See PARALLEL RULES.

wind surfing

hang gliding underwater.

wind vane

a small flag, PENNANT or square of light metal (for durability) used as a means of indicating the direction of the wind and the way it is shifting; usually rigged at the MASTHEAD but sometimes seen attached to the SHROUDS of a racing boat.

windward

the side of a sailing vessel nearest the wind; if in doubt, you usually have to climb uphill to get there. See also LEEWARD.

wing-on-wing

see GOOSE-WINGED.

wishbone spar

a spar shaped somewhat like a chicken wishbone, the ends of its two halves extending the CLEW of a LOOSE-FOOTED sail; this system is used on Freedom yachts so that the mast can be rotated to reef the sail. See UNSTRESSED RIGS.

wooden boats

an increasingly rare sight in the world of ocean cruising, but there is something magical about treading the DECKs of old, timber-built craft—as Ishmael recalled when describing the *Pequod*.

> You may have seen many a quaint craft in your day, for aught I know;—square-toed luggers; mountainous Japanese junks; butter-box galliots, and what not; but take my word for it, you never saw such a rare old craft as this same rare old *Pequod*. She was a ship of the old school, rather small if anything; with an old-fashioned claw-footed look about her. Long seasoned and weather-stained in the

typhoons and calms of all four oceans, her old hull's complexion was darkened like a French grenadier's, who has alike fought in Egypt and Siberia. Her venerable bows looked bearded. Her masts—cut somewhere on the coast of Japan, where her original ones were lost overboard in a gale—her masts stood stiffly up like the spines of the three old kings of Cologne. Her ancient decks were worn and wrinkled, like the pilgrim-worshipped flag-stone in Canterbury Cathedral where Becket bled.

Moby Dick; or The Whale by Herman Melville, 1851

working sails

the sails you hope you do not ever have to change; MAINSAIL, MIZZEN, and JIB (or ROLLER-REEFED GENOA).

wrecks

For those of you still unconvinced by my admonitions about the BERMUDA TRIANGLE and the MARY CELESTE, this spooky tale dates from 1757.

Then, it was nigh on to evening, we came upon a creek opening into the greater one through the bank upon our left. We had been like to pass it—as, indeed, we had passed many throughout the day—but that the bo'sun, whose boat had the lead, cried out that there was some craft lying-up, a little beyond the first bend. And, indeed, so it seemed; for one of the masts of her—all jagged, where it had carried away—stuck up plain to our view.

Now, having grown sick with so much lonesomeness, and being in fear of the approaching night, we gave out something near to a cheer, which, however, the bo'sun silenced, having no knowledge of those who might occupy the stranger. And so, in silence, the bo'sun turned his craft toward the creek, whereat we followed, taking heed to keep quietness, and working the oars warily. So, in a little, we came to the shoulder of the bend, and had plain sight of the vessel some little way beyond us. From the distance she had no appearance of being inhabited; so that after some small hesitation, we pulled towards her, though still being at pains to keep silence.

The strange vessel lay against that bank of the creek which was upon our right, and over above her was a thick clump of the stunted trees. For the rest, she appeared to be firmly imbedded in the heavy mud, and there was a certain look of age about her which carried to me a doleful suggestion that we should find naught aboard of her fit for an honest stomach.

We had come to a distance of maybe some ten fathoms from her starboard bow—for she lay with her head down towards the mouth of the little creek—when the bo'sun bade his men to back water, the which Josh did regarding our own boat. Then, being ready to fly if we had been in danger, the bo'sun hailed the stranger; but got no reply, save that some echo of his shout seemed to come back at us.

And so he sung out again to her, chance there might be some below decks who had not caught his first hail; but, for the second time, no answer came to us, save the low echo—naught, but that the silent trees took on a little quivering, as though his voice had shaken them.

At that, being confident now within our minds, we laid alongside, and, in a minute had shinned up the oars and so gained her decks. Here, save that the glass of the skylight of the main cabin had been broken, and some portion of the framework shattered, there was no extraordinary litter; so that it appeared to us as though she had been no great while abandoned.

So soon as the bo'sun had made his way up from the boat, he turned aft toward the scuttle, the rest of us following. We found the leaf of the scuttle pulled forward to within an inch of closing, and so much effort did it require of us to push it back, that we had immediate evidence of a considerable time since any had gone down that way.

However, it was no great while before we were below, and here we found the main cabin to be empty, save for the bare furnishings. From it there opened off two state-rooms at the forrard end, and the captain's cabin in the after part, and in all of these we found matters of clothing and sundries such as proved that the vessel had been deserted apparently in haste. In further proof of this we found, in a drawer in the captain's room, a considerable quantity of loose gold, the which it was not to be supposed would have been left by the free-will of the owner.

Of the state-rooms, the one upon the starboard side gave evidence that it had been occupied by a woman—no doubt a passenger. The other, in which there were two bunks, had been shared, so far as we could have any certainty, by a couple of young men; and this we gathered by observation of various garments which were scattered carelessly about. Yet it must not be supposed that we spent any great time in the cabins; for we were pressed for food, and made haste—under the directing of the bo'sun—to discover if the hulk held victuals whereby we might be kept alive.

To this end, we removed the hatch which led down to the lazarette, and, lighting two lamps which we had with us in the boats, went down to make a search. And so, in a little while, we came upon two casks which the bo'sun broke open with a hatchet. These casks were sound and tight, and in them was ship's biscuit, very good and fit for food. At this, as may be imagined, we felt eased in our minds, knowing that there was no immediate fear of starvation. Following this, we found a barrel of molasses; a cask of rum; some cases of dried fruit—these were mouldy and scarce fit to be eaten; a cask of salt beef, another of pork; a small barrel of vinegar; a case of brandy; two barrels of flour—one of which proved to be damp-struck; and a bunch of tallow dips.

In a little while we had all these things up in the big cabin, so that we might come at them the better to make choice of that which was fit for our stomachs, and that which was otherwise. Meantime, whilst the bo'sun overhauled these matters, Josh called a couple of the men, and went on deck to bring up the gear from the boats, for it had been decided that we should pass the night aboard the hulk.

When this was accomplished, Josh took a walk forward to the fo'cas'le; but found nothing beyond two seamen's chests; a sea-bag, and some odd gear. There were, indeed, no more than ten bunks in the place; for she was but a small brig, and had no call for a great crowd. Yet Josh was more than a little puzzled to know what had come to the odd chests; for it was not to be supposed that there had been no more than two—and a sea-bag—among ten men. But to this, at that time, he had no answer, and so, being sharp for supper, made a return to the deck, and thence to the main cabin.

The Boats of the "Glen Carrig"
by William Hope Hodgson, 1757

Of course there are a number of possible explanations for this early version of the MARY CELESTE, but the most likely has to be the late-night swimming without either (a) checking that you can get back onboard or (b) that the wind is not rising.

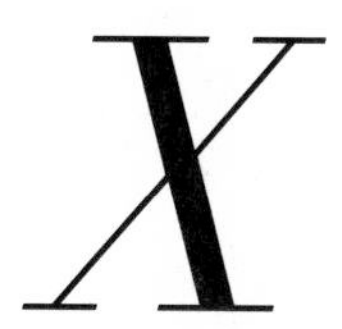

X-Files

see BERMUDA TRIANGLE.

X-tant

generic term for a navigation instrument used to measure the ELEVATION of a CELESTIAL OBJECT; QUADRANT, SEXTANT, er, ASTROLABE?

XTE

abbreviation for CROSS-TRACK ERROR.

Y

yacht

a decked sailing boat; generic term for a boat powered by sail. (In the US, the more common term is SAILBOAT; used to distinguish from a POWERBOAT.) In everyday parlance, "yacht" is used as shorthand for "luxury power cruising yacht." When in 1980 Robert Maxwell (former British member of parliament, business tycoon, and publisher) said that he "came from a farm-labouring family and [didn't] go in for owning yachts or going to big parties," he didn't mean he couldn't afford a Wayfarer. In 1991, while under investigation for fraud, he "fell off" his "yacht," the *Lady Ghislaine*, and drowned. Maxwell owned the vessel, which did not sport a single sail and was in any case much bigger than a Wayfarer.

In 1964, only a few months after the assassination of President John F. Kennedy, President Charles de Gaulle of France said this about Mrs. Jacqueline Kennedy: "I can see her in about ten years from now on the yacht of a Greek oil millionaire." Guess what happened next? You couldn't make it up. (And that "yacht" wasn't a Wayfarer, either.)

"Yammerschooner!"

see FIREARMS.

Yankee

a large triangular HEADSAIL used in light winds.

yard

a SPAR that extends horizontally across a mast in order to extend a working SQUARE SAIL; not to be confused with a BOOM, GAFF, or SPRIT, which extend FORE-AND-AFT. See WHOLE NINE YARDS.

yardarm

the part of a YARD which extends beyond the top corner of a SQUARE SAIL (the EARING); device which used to double-up as a gallows so that offenders convicted onboard ship could be hanged.

> Whereas, William White, Boatswain's mate of HMS *Gibraltar*, but serving on board HMS *Courageux*, hath been tried by a Court Martial on charges exhibited against him by Lt John Glover, for having seized him by the collar, and wishing he had the said Lt Glover on shore, he would then do his business, and other mutinous expressions; and the Court having found him guilty of the 22nd Article of the Articles of War, hath adjudged the said William White to suffer death by being hanged by the neck at the yard-arm of such Ship, as the Commanding officer for the time being might direct.
>
> Admiral Horatio Nelson, 1799

This, of course, in days when we were so unsophisticated that we judicially killed people for seizing someone by the collar. On second thoughts, however, it could be brought back for certain shipboard offenses such as (a) failure of crew to brew up for next watch, (b) anyone putting something down the HEAD(S) that has not just been all the way through their alimentary canal or (c) the next person to lose a FENDER over the side. Three Strikes and You're Out. See also KEEL-HAULING.

And, of course, when "the sun is over the yardarm" it is time for the first drink of the day.

yaw, to

to swing from one side of the intended course to another. Causes can vary (e.g., a shifting, light following wind while running); this situation demands good concentration and anticipation by the helmsman.

yawl

a twin-masted FORE-AND-AFT rigged yacht not unlike a KETCH, but with the smaller, sternmost mast set *aft* of the RUDDER POST.

yellow fever

see MOSQUITO.

yuloh

a Chinese SCULL superior in design to those seen in the West. A yuloh is curved or jointed just beyond the TRANSOM in such a way that the aft section curves down toward the water giving the blade a better purchase. *Yu* is Chinese for fish and, I think, *yuloh* means "fishtail," which is very descriptive of the function it performs and how it does it.

Z

zenith

in ASTRO-NAVIGATION (nautical astronomy) the point on the CELESTIAL SPHERE immediately above the OBSERVER (i.e., if a straight line is drawn from the center of the earth, through the observer, it will touch the celestial sphere at the observer's zenith). (It will touch his or her soul, too, because it probably means they can navigate without GPS.)

Zheng He

Zheng He (pron. "jung her") was commander of China's great Imperial Navy in the early 15th century. It has already been well documented that one of Zheng's fleets sailed out of the South China Sea and across the Indian Ocean to reach the east coast of Africa. But recent evidence suggests (but does not prove) that he also led an expedition across the Pacific and fetched up at Baja California seventy-two years before Columbus discovered that the Bahamas has a lot of reefs.

This interesting research was carried out by Gavin Menzies, a retired Royal Navy submarine commander who has also produced evidence that Magellan may have been in possession of a map showing South America, Africa, and Australia. Venetian explorer Nicolo da Conti, who supposedly once sailed with Admiral Zheng, "borrowed" the map from the Chinese and then, presumably, sold it to the Portuguese. Commander Menzies's book, *1421: The Year China Discovered the World*, was published in 2003 and became a bestseller.

Sadly, though, *1421* turned out to be the exploration/seafaring equivalent of *The Da Vinci Code*: it's hokum. The Wikipedia entry for the book describes at length the extensive academic challenges to Menzies's research. In my book, *The Barefoot Navigator*, I describe why his claimed Chinese method of determining longitude using lunar eclipses could not work. (To declare an interest, I speak Chinese and have spent a lot of time in China; so there is no animosity on that front.)

Undeterred, Menzies went on to claim in a subsequent book that, not only did the Chinese discover the whole world, they were also responsible for the Italian Renaissance. Could they have originated the da Vinci code?

zinc anodes

small ingots of zinc attached to the hull of a metal boat to help protect it from corrosion due to electrolysis; in practice, on a steel boat, more likely to be great big limpet mines fixed to the hull, keel, and skeg. Electrolysis is an effect caused when dissimilar metals are placed in a saltwater solution (the sea, for example). This problem is particularly acute in some marinas and harbors where shore supply power cables are left dangling in the water.

zulu

1. military shorthand for GREENWICH MEAN TIME; so, 1400Zulu = 1400 GMT or even 1400UT (Universal Time). The use of the letter originates from the designations given to time zones. The Greenwich time zone is Z. Each zone going to the east is allocated the letters A to I, then K, L, and M; J is omitted. Each zone going west from Greenwich is allocated N to X Y. M and Y designate each side of the INTERNATIONAL DATELINE within the ± 12 hour zone.
2. Scottish fishing boat.

SELECTED BIBLIOGRAPHY

The following is just a few of the books I referred to during the compilation of this lexicon. If no publication details are shown, the book may be out of print and you will have to resort to a good library for a reference copy. If the book is now in the public domain you may be able to find a free download copy on the Internet via the usual search engines. Some popular classics are, of course, in the public domain but remain in print. Online booksellers such as Amazon.com are a good way of sourcing these.

Ayto, John. *Dictionary of Word Origins*. London: Bloomsbury, 1990–1991.

Beatty, John. *The Breath of Angels*. London: Mainstream, 1995; New York: Sheridan House, 1997.

Blythe, Donna G. "The Ciguatera Homepage." www.rehablink.com/ciguatera.

Boteler, Nathaniel. *Six Dialogues about Sea Services*. London: Moses Pitt, 1685.

Bowditch, Nathaniel. *The American Practical Navigator*. Bethesda, MD: National Imagery and Mapping Agency, 2002.

Brewer, Ebenezer Cobham. *The Brewer Dictionary of Phrase and Fable*. Originally published 1870; various updated editions available.

Browne, J. Ross. *Etchings of a Whaling Cruise*. New York: Harper & Brothers, 1846.

Carson, Rachel. *The Sea Around Us*. London: Oxford University Press, 1951.

Chaucer, Geoffrey. *Treatise on the Astrolabe*. c.1391; various sources including the Internet.

Childers, Erskine. *The Riddle of the Sands*. London: Seafarer Books; New York: Sheridan House, 1998; originally published 1903.

Conrad, Joseph. *The Mirror of the Sea*. New York: Marlboro, 1988; originally published 1903.

Cook, Capt. James. *The Journal of Captain Cook*. New York: Penguin, 2000; originally published 1773.

Dallas, Sir George. *The India Guide; or, Journal of a Voyage to the East Indies in the Year MDCCLXXX*. Calcutta, 1785.

Dana, Richard Henry Jr. *Two Years Before the Mast: A Personal Narrative*. New York: Signet, 2000; originally published 1840.

Darwin, Charles. *The Voyage of the Beagle*. New York: Narrative Press, 2001; originally published 1845.

Davis, Captain John. *The Seaman's Secrets*. 1585.

Defoe, Daniel. *Robinson Crusoe*. Princeton: Princeton Review, 2001; originally published 1719.

De Rougemont, Louis. *The Adventures of Louis De Rougemont*. by Himself (1898?); Indypublish.Com, 2001.

Drury, Robert. *Madagascar: or Robert Drury's Journal During 15 Years' Captivity on that Island*. Westport, CT: Greenwood Publishing Group, 1970; originally published 1728.

Golding, Sir William. *Rites of Passage*. New York: Farrar, Straus & Giroux, 1980.

Hanson, Neal. *The Custom of the Sea*. London: Doubleday, 1999.

Hayden, Sterling. *Wanderer*. New York: Sheridan House, 1998.

Hodgson, William Hope. *The Boats of the "Glen Carrig."* New York: Wildside Press, 2001; originally published 1757.

Homer. *The Odyssey*. New York: Noonday Press, 1998; originally published 800 BC.

Knight, E. F. *Cruise of the "Alerte" in Search of Treasure*. New York: Narrative Press, 2002; originally published 1890.

———. *The Falcon on the Baltic: A Coasting Voyage from Hammersmith to Copenhagen in 1887*. London: W. H. Allen, 1892; New York: Narrative Press, 2003.

Knights, Captain Arthur E. *Notes by the Way in a Sailor's Life*. 1898.

Lagan, Jack. *The Barefoot Navigator*. London: Adlard Coles Nautical; New York: Sheridan House, 2006.

Laird, Edgar. "Astrolabes and the Construction of Time in the Late Middle Ages." In *Constructions of Time in the Late Middle Ages*, edited by Carol Poster and Richard Utz. Evanston, IL: Northwestern University Press, 1997.

London, Jack. *The Cruise of the Snark*. London: Seafarer Books; New York: Sheridan House, 2000.

———. *Dutch Courage and Other Stories*. Des Moines, IA: Classic Books, 2000; originally published 1901.

———. *The Human Drift*. Indypublish.Com, 2002; originally published 1911.

Mackinnon, A. J. *The Unlikely Voyage of Jack de Crow*. London: Seafarer Books; New York: Sheridan House, 2002.

Melville, Herman. *Moby Dick; Or The Whale*. New York: Bantam Classics, 1981; originally published 1851.

Menzies, Gavin. *1421: The Year China Discovered the World*. New York: Morrow, 2003.

Mielche, Hakon. *Let's See If The World Is Round*. London: William Hodge & Co. Ltd., 1944; originally published in Danish as *Monsunens Siste Rejse*, 1938.

Pardey, Lin, and Larry Pardey. *Storm Tactics Handbook—Modern Methods of Heaving-To for Survival in Extreme Conditions*. Arcata, CA: Pardey Books; Ramsbury, UK: Airlife Publishing Ltd., 1996.

Peniakoff, Vladimir. *Popski's Private Army*. London: Cassell Military Paperbacks, 2002; originally published 1950.

Perkins, Roger, and Captain K. J. Douglas-Morris. *Gunfire in Barbary*. London: Kenneth Mason, 1982.

Pretty, Francis. *Sir Francis Drake's Famous Voyage Round the World*. 1577.

Raban, Jonathan. *Coasting*. London: Collins Harvill Ltd., 1986.

Ransome, Arthur. *Racundra's First Cruise*. London: David & Charles, 1988; originally published 1923. (I have noticed a reference to *Racundra's Third Cruise*, so I assume there must be a *Racundra's Second Cruise* also waiting to be found.)

Roberts, Bob. *Coasting Bargemaster*. London: Seafarer Books, 2000.

———. *Last of the Sailormen*. London: Seafarer Books, 2002.

Robertson, Dougal. *Survive the Savage Sea*. London: Elek Books, 1973; New York: Sheridan House, 1994.

Rose, Sir Alec. *My Lively Lady*. London: Nautical Publishing Co. and George C. Harrap & Co. Ltd., 1968; London: Pan Books Ltd., 1969; Nautical Publishing Co. (reprint date uncertain).

Slocum, Capt. Joshua. *Sailing Alone Around the World*. New York: Sheridan House; London: Seafarer Books, 1993; originally published 1896.

———. *Voyage of the "Liberdade."* New York: Narrative Press, 2001; originally published 1894.

Smith, Tony. *The BMA Complete Family Health Guide*. London: Dorling Kindersley, 1999.

Sobel, Dava. *Longitude*. New York: Walker Publishing, 1995; London: Fourth Estate, 1996.

Stadler, Michael. *Psychology of Sailing*. London: Adlard Coles, 1987.

Stevenson, Robert Louis. *The Wrecker*. Gainsville, FL: Blue Unicorn Editions, 1997; originally published Edinburgh, 1896.

Thomas, Donald. *Cochrane: Britannia's Sea Wolf*. London: Cassell, 2001; originally published by André Deutsch, 1978.

Thomson, George Malcolm. *Sir Francis Drake*. London: Secker & Warburg Ltd., 1972.

Walton, Izaak. *The Compleat Angler*. London: Everyman Paperback Classics, 1993; originally published 1653.

Willis, William. *The Epic Voyage of the Seven Little Sisters*. London: Hutchinson & Co. Ltd., 1956.